D0919438
WINCHESTER, VA.

THE STEAM ENGINE OF THOMAS NEWCOMEN

The Steam Engine of Thomas Newcomen

L.T.C. Rolt & J.S. Allen

Numine Nitor

Moorland Publishing Company
Hartington
Science History Publications/USA
New York – 1977

Library of Congress Cataloging in Publication Data

Rolt, Lionel Thomas Caswell, 1910-
The steam engine of Thomas Newcomen.

Bibliography: p.
1. Steam-engines—History. 2. Newcomen, Thomas, 1663-1729. I. Allen, John S., joint author. II. Title.
TJ485.N48R64 1977 621.1'64 77-22930
ISBN 0-88202-171-0

ISBN 0 903485 42 7

Published in Great Britain by
Moorland Publishing Company,
The Market Place, Hartington,
Buxton, Derbys, SK17 0AL

First published in the United States by
Science History Publications/USA
a division of
Neale Watson Academic Publications, Inc.
156 Fifth Avenue, New York 10010

Printed in Great Britain by
Wood Mitchell & Co Ltd, Stoke on Trent.

Contents

Illustrations

Lionel Thomas Caswell Rolt 1910-74

L.T.C. Rolt, the engineering and transport historian gained public acclaim through his writings and his active involvement in many projects concerned with the preservation of industrial history. His work on industrial archaeology helped to create a growing public interest in the subject and all his books reflected his love of good engineering, the English language and the English countryside.

He had a wide industrial training as a mechanical engineer and from his work in small engineering companies sprang not only his love of machines, but a deep understanding of the human side of industrial enterprise. One can learn much of Rolt's youth and early manhood from *Winterstoke* and *Landscape with Machines.*

Following a period in which he was concerned with vintage motor cars he acquired a narrow boat and voyaged over much of the canal system of the Midlands. It was this experience which he described so richly in *Narrow Boat.* This was the pathfinder for the whole post-war revival of interest in canals for pleasure purposes.

His infectious zeal then involved him in narrow-gauge railway preservation and restoration and in particular the Tal-y-Llyn Railway. He became general manager of the line during and after its restoration. Through this and earlier activities he gained first-hand experience of the work of past engineers, so that when he wrote his biographies of *James Watt, Thomas Telford, George and Robert Stephenson, I.K. Brunel* and his many engineering histories he was able to bring to the study of his subjects a direct knowledge of materials, men and machines.

Following Rolt's untimely death the Council of the Newcomen Society for the Study of the History of Engineering and Technology agreed unanimously to initiate some form of permanent memorial. Considered to be engineering biography at its best, Rolt's *Thomas Newcomen, The prehistory of the Steam Engine,* was published in 1963 as part of the commemoration of the 300th anniversary of the date of Newcomen's birth. The Society decided that a revised and enlarged memorial edition would be a most fitting tribute. J.S. Allen, a member of Council and Chairman of the Midlands Branch, who had made a long and close study of Thomas Newcomen and his work consented to prepare the book, adding to it in a singularly effective way the results of his own enthusiasm and the fruits of his researches.

The Council of the Newcomen Society welcomes this book as a fitting memorial and a token of its regard for L.T.C. Rolt who was Vice-President of the Society.

Wilson of High Wray
President of the Newcomen Society 1974-6
Acting President 1977

The Rolt Memorial

The Council of the Newcomen Society for the Study of the History of Science and Technology commissioned the present work as a Memorial volume to L.T.C. Rolt. They formed the Rolt Memorial Sub-Committee whose members were Lord Wilson of High Wray, (Chairman and President of the Newcomen Society 1974-5, 1975-6, Acting President 1977), Mr F. Nixon, (President of Newcomen Society 1976-7), Professor A. Skempton, Dr A. Buchanan, Messrs J.S. Allen, J.W. Butler, L.R. Day, E.C.R. Hadfield, R.J. Law, T.M. Simmons, J.L.E. Smith.

Mrs Sonia Rolt agreed that a revised and enlarged edition of her husband's book would be a suitable memorial to him and she willingly gave her consent to its use. The more so, in that John Allen has given the original book the results of his researches over a number of years and she expressed the view that L.T.C. Rolt himself would have welcomed the energetic and productive research now added.

Acknowledgements

L.T.C. Rolt in the acknowledgements to his book on Thomas Newcomen in 1963 referred to the fact that it was based partly on the work of early writers on the history of the steam engine and partly on the fruits of the immense amount of research work which had been done during this century. In this research the late Dr H.W. Dickinson and the late Rhys Jenkins had led the way and the book owed a great deal to their inspiration and scholarhsip. He was also deeply indebted to those members of the Newcomen Society who had followed their example and whose writings will be found listed in the bibliography. Dr Raistrick's work had proved particularly valuable. He was particularly indebted to Mr A. Stowers for many helpful comments and to Mr Percy Russell of Dartmouth for much kindness and for so freely imparting the results of his local researches.

Throughout the years of his researches the present writer has had the assistance of very many people. They have been separately acknowleged in the several papers published by the Newcomen Society but this opportunity is taken to thank them again for their help and advice. Reference to all published books is included in the text and with some exceptions other information is taken from the papers and sources listed in the bibliography. To those interested the papers published by the Newcomen Society contain fully detailed references to all primary sources. Unpublished sources used include information from the Swedish Archives on Marten Triewald and the author is particularly indebted to Professor Althin and Mr Svante Lindqvist of the Royal Institute of Technology, Sweden, for their co-operation and assistance with this. In addition Mr Lindqvist generously provided information from his researches on John O'Kelly and from the diary of the Swedish traveller Kalmeter.

Mr Torsten Berg has provided a new translation of Triewald's book *A Short Description of the Fire and Air Engine,* and also translated sections of the diary of Jonas Alstromer. Information on the Proprietors of the Patent and other advice has been provided by Mr Alan Smith, now of the British Embassy, Washington, USA. Mr J.V. Beckett of Fairleigh Dickinson University, Banbury, has provided details of the later Whitehaven engines, reference to the original partnership for the Ginns engine, and the financial affairs of the Proprietors from the Lowther papers at Carlisle. Mrs S.B. Rogoyska kindly loaned the Newcomen family papers in her possession.

I am grateful to Mr J.S. Roper, of Woodsetton, Dudley for his encouragement and assistance in a number of detail points. Mr E. Grindley of the Friends of the Dartmouth Museum Association has been most helpful with illustrations and information. Others have helped with the preparation of the book itself and in this connection my sincere thanks must first be extended to Professor J.R. Harris of the Department of Economic History, The University of Birmingham, who studied the draft in detail and made many helpful contributions and comments. I am most grateful for all the assistance that he has given in this and many other ways.

Mr R.J. Law of the Science Museum, London has been most helpful in the search for illustrative material. Dr G. Hollister-Short of Shoreditch College, Egham, has allowed me to use his recent thesis on the introduction of the engine into Europe and kindly commented on the draft of that section. The Librarian of the Institution of Civil Engineers kindly arranged for the loan of early published works in his care. It must be rare indeed for an author to have as publisher, an expert in the particular subject of the book. Such has been my good fortune and I am greatly indebted to Dr J.A. Robey, of the Moorland Publishing Company for his help, particularly with information on the later engines and in the preparation of the plans. He also gave permission for the use of a unique list prepared by himself and Mr J. Kanefsky which records details of all known engines built during the eighteenth century. Mrs Sonia Rolt kindly read the proofs and gave much helpful assistance.

I must express my sincere appreciation to Mrs Barbara Johnson who so kindly typed the drafts.

I wish to express thanks to the following who supplied illustrations and granted reproduction permission:

The Director of the Science Museum, South Kensington, Figs 1-10, 17-19, 25, 27, 34, 37-45, 48-50, 55-9, 61, 62, 65, 67, 70, 72, 77, 80-2, 84-7, 90-3; The Institution of Civil Engineers, Figs 12, 21, 52, 53, 66, 95, 96; The Institution of Mechanical Engineers, Figs 63, 75, 83, 97; Bodleian Library, Oxford, Figs 22, 23; Mr Alan Smith and Mr Bernard Gilonis Fig 24; Dipl Ing A. Hoffman, Kassel, with acknowledgements to the estate of Dipl Ing Dietrich Hoffman, Figs 28, 29, 32; British Museum 30, 32; National Archives of Hungary, Collection of Plans, Fig 33; Mr Svante Lindqvist, Royal Institute of Technology Library, Stockholm, Fig 35; Newcomen Society, Fig 36; Mr J.G. James, Fig 11; Rev R.S. Chalk, Stoke Fleming, Figs 13, 14; Devon Record Office, Exeter, Fig 15; Staatsbibliothek, Berlin, Fig 46; Dartmouth Borough Museum, Fig 47; County Archivist, Northumberland Record Office, Newcastle on Tyne, Fig 51; Sheffield City Libraries, Fig 54; Mr Frank Nixon, Figs 69, 73, 74; The Henry Ford Museum, Dearborn, USA, Figs 88, 89; Dr C.T.G.Boucher, Fig 94; Dr J.A. Robey and Mr J. Kanefsky, Fig 99; Mr R.H. Bird, Fig 60; Mr F.D. Woodall, Fig 64; National Coal Board, Figs 76, 78-9; Dr J.A. Robey assisted in the production of the plans and diagrams Figs 20, 23a, 26, 98, 99.

J.S. Allen

A Doctor sage, new coming in,
Condemn'd the Methods that were us'd before;
And said,—That I in Caves shou'd dwell no more:
Then I shou'd dwell in free and open Air,
And gain new Vigour from the Atmosphere:
An House for me he built—Did Orders give,
I shou'd no Weight above my Strength receive;
And that I shou'd, for Breath, and Health to guard,
Look out of Windows when I labour'd hard.
These gentle Means my Shape have alter'd quite;
I'm now encreas'd in Strength, and Bulk, and Height;
I can now raise my Hand above my Head;
And now, at last, I by my self am fed.
On mighty Arms, alternately I bear
Prodigious Weights of Water and of Air;
And yet you'll stop my Motion with a Hair

—Henry Beighton

The Prize AEnigma, 1725

Introduction

This book has been written to honour the name and the achievement of a man who was born three hundred years ago. His bones lie neglected in a London burying-ground — their very place unknown. He brought fame to his town, his county and his country, and to the whole world he gave the art of converting fuel into useful power for the benefit and convenience of humanity.

These words are a paraphrase of those spoken by Loughnan St L. Pendred, then President of the Newcomen Society for the Study of the History of Engineering and Technology, when the bicentenary in 1929 of the death of Thomas Newcomen was commemorated at Dartmouth, Devon, the place of his birth. Mr Pendred went on to speak of the obscurity which surrounded the name and the life of this great engineer and what he said is still just as true today. Members of the society which proudly bears Newcomen's name can testify from first-hand experience how totally this name has been eclipsed by that of James Watt in popular esteem. For how often does mention of the society evoke a puzzled expression and a question 'Newcomen, what does that mean?'

To answer such questions as this, is the object of this book. It has been written with the active help, advice and encouragement of a society which has always been concerned to win for Newcomen the acclaim that his genius deserves. For over two centuries the fame of Watt has eclipsed that of Newcomen. The work of scholars on the subject over the last half century, was led in the early thirties by Rhys Jenkins who paved the way with a series of papers on Savery, Newcomen, Henry Beighton and the early history of the steam engine.

More recently others have taken up the topic just as assiduously and particularly fruitful new sources including legal and estate records have been studied and the results of this work now demand that Newcomen should have a much greater share of the credit. It was he who first produced a machine capable of providing power, other than that derived from man, animals or the elements and furthermore, providing it with technical efficiency and regularity, and on acceptable commercial terms.

The evidence which would have shown the extent of his success was not discovered, mainly because earlier historians did not look for it. Once it is realised that over a hundred of Newcomen's engines had been set up during the period of the patent, which expired only four years after his death (a number which may yet well be increased) and that at least 1,500 of his engines (almost certainly far more) were built during the eighteenth century, the picture is transformed. Even during the period of the Watt patent (1769-1800) the number of Newcomen engines erected exceeded those built by Watt. Considering the difficulties of exporting the technology and the engine's dependence on large quantities of cheap fuel, the European interest taken during its first decades was remarkable, and by the middle of the eighteenth century a considerable number of descriptions, drawings and technical explanations had been published at home or abroad, often in learned works. The early engines cost over £1,000 each to instal and also required a licence, thus they were never a cheap item of capital equipment, and were, for the most part, confined to the various branches of the mining industry. Given these limitations, their rate of diffusion was truly amazing and the common belief that steam engines were only taken up in limited numbers before Watt's time is grotesque. Rather than supposing the Newcomen engine to be an example of slow technical take-up we should be asking if there is any item of early eighteenth-century technology requiring a significant capital expenditure which had a rate of adoption which is remotely comparable.

By founding the Royal Society in 1660, Charles II set the seal of royal patronage and approval upon the new spirit of scientific inquiry and speculation which, in England as in Europe, was the *zeitgeist* of the Renaissance. Yet this significant ferment of activity was, in Newcomen's day, still almost exclusively confined to the study and the laboratory bench.

Knowledge was pursued for its own sake and, although very significant discoveries were made, the discoverers were seldom concerned to apply their knowledge to practical purpose. In other words 'Natural Philosophy', as it was then called, was the equivalent of what we now term pure science. For the instruments and apparatus they needed, the seventeenth-century scientists turned to the great horologists of the day and until the century ended the beautiful work which these master craftsmen executed in this new sphere remained the only practical evidence that a new spirit was abroad.

Between this small and exclusive band of craftsmen and the makers of all those practical things by which man seeks to lighten his daily toil an immense gulf was fixed. For example, the makers of wind and water mills built solidly and well, but with such massive crudity and ignorance of principle that it is almost impossible for us to believe that workmanship so medieval in character could be contemporary with that of the great horologists and instrument makers of seventeenth-century London. For many years there was no bridging this gulf. Instrument making was a 'mystery', a jealously guarded closed shop, while the scientists of the day looked upon the practical man, who lived by the skill of hand and eye, with a certain arrogance and contempt that is typified in their attitude towards Thomas Newcomen.

If there is one fact certain about Newcomen it is that he belonged to the ranks of the practical men. He was a provincial tradesman and, to the small scientific world of London whose centre was the Royal Society, it seemed inconceivable that such a man could excel the brightest intellects of the day in power of invention. So the majority ignored Newcomen and his engine completely, while the few who deigned to mention his achievement did their best to belittle it and to explain it away. They stressed Newcomen's indebtedness to earlier scientific experiments and many of the original ingenious details to fortunate chance, to the help of others more learned and skilled than himself, or to the direct intervention of the Almighty. Anything rather than admit that so potent a machine could be the child of Newcomen's own genius and unwearying perseverance.

At the time of its conception Newcomen's engine was an uniquely successful marriage between new scientific principles and practical engineering as practised by the mill-wrights of the day. The genius of its inventor bridged the gulf which had hitherto precluded such a union, and in the long term its effect was tremendous. Invention begets invention. The very imperfections of Newcomen's engine, the crudity of its construction and, most notably, that insatiable appetite for fuel which was the symptom of its very low thermal efficiency, were a technical challenge which no ingenious man could resist. By the middle of the eighteenth century a very great number of Newcomen engines were in use and this had a twofold effect. The range of the engine's influence as a stimulus to further invention was greatly extended, while the practical results of its widespread employment as a colliery pump led directly to those practical improvements in manufacturing technique which alone made the realization of new inventions possible. More effective colliery pumping spelled more coal; more coal spelled more iron and improved methods of iron production. A rapidly developing iron industry could supply Newcomen's successors with cast or forged iron components of a size and quality such as he had never known.

Watt's separate condenser and his double-acting rotative steam engine, which proved so versatile in application, were undoubtedly the greatest advance in mechanical engineering of the late eighteenth century. The new engine was hailed by a world that had forgotten Newcomen as a giant's stride forward on the road to a scientific Utopia and its inventor was lauded as a universal benefactor. So originated the still current popular fallacy that James Watt 'invented' the steam engine.

It is no purpose of this book to belittle Watt's achievement — his fame was richly deserved — but rather to pay honour where honour is due. In any new field of human endeavour the first step forward is always the most difficult and most venturesome. In the story of the development of mechanical power from heat it was Thomas Newcomen who took that first momentous step. He was the true pioneer and all who followed after him were but great improvers, building upon the firm but forgotten foundation which he had laid.

An explanatory word must now be said about the term 'steam engine' which has been freely used here and throughout this book. It has been so used for the sake of simplicity, but in its modern sense of an engine which derives its power from the pressure and expansive force of steam the term is misleading when applied to the engines of Newcomen and Watt. The power of steam was known to both men, but because of the technological limitations of their age they harnessed another power — 'the weight of the atmosphere' — and used steam merely as the most convenient agent for the creation of a vacuum so that they could bring this power to bear. In strict terminology Newcomen's invention should be called an atmospheric engine because it was the pressure of air that forced its piston down the cylinder when a vacuum had been created below it. In the pursuit of higher thermal efficiency James Watt substituted hot steam for cold air in his cylinder, but it is too often forgotten that it was still the creation of a vacuum below the piston which gave that steam the power to act. Its pressure exceeded that of the earth's atmosphere by so small a margin that it would otherwise have been powerless.

Nevertheless, despite this vital difference in working principle the first Newcomen engine, with its combination of boiler, cylinder, piston and automatic valve gear is, as a machine, the undoubted sire of the steam engine of modern times. Like Watt's refinements, the introduction of high-pressure steam was accomplished without making any fundamental change in those major components which the genius of Newcomen first successfully combined. In such a book as this, therefore, to refer to Newcomen's invention as anything other than a steam engine would be sheer pedantry.

Finally, if this book were to be strictly confined to the lifetime of Newcomen it would be relatively short. Instead, it begins with a review of the unsuccessful attempts which were made to harness steam or atmospheric power before Newcomen's day, while a final section takes the history of his engine forward to the end of the eighteenth century. Such a prelude and postlude are essential to any assessment of his achievement. In this respect the history of the Newcomen engine after its inventor's death is particularly important because one result of the eclipse of Newcomen by Watt has been that the development of the former's engine by succeeding engineers, its widespread popularity and its stubborn persistence even in the face of competition from the more economical Watt type are facts which have never been sufficiently widely recognized. They need stressing. The pages of the history of invention are filled with 'might-have-beens', men who tried to run before they had learned to walk; men whose ambitious schemes may seem far-sighted in retrospect but were in their day utterly impracticable because they were not tailored to current needs or resources. The man this book celebrates was emphatically not one of this unsuccessful and frustrated company. He was the first great mechnical engineer and his successful creation was of incalculable importance, not only in the history of his own country but of the world. His memory will live as long as history is written.

CHAPTER 1

The Progenitors

Early Attempts to Harness Steam Power

One of the least attractive aspects of the rapid progress of science and technology during the three centuries which have passed since Newcomen was born is the intellectual arrogance which it has bred. This *hubris*, as the Greeks called it, is particularly evident in our attitude towards the past. Because we wrongly equate the possession of the material powers which scientist and engineer have placed in our hands with civilization, we are too apt to assume that those who commanded none of these powers must necessarily be our inferiors in intellectual capacity. Despite the existence of a cloud of witnesses in this country alone from Chaucer to Milton, an obsession with technology blinds us to the truth that it is not the only path which the questing mind of man can pursue. Our forerunners possessed intellectual powers equal or superior to our own but, because their conception of the world and of man's place in it was very different from ours, they applied those powers to other ends. The European Renaissance did not bring about any sudden, mysterious increase in man's capacity. What it did was to change his attitude towards his world. A new 'world picture' concentrated man's powers upon scientific inquiry as a lens focuses the heat of the sun.

An excellent example of our wrong thinking about the past is the stubborn persistence of the legend that the youthful James Watt 'discovered' the power of steam by observing the lifting lid of a kettle as it boiled on the hearth of his home in Greenock. A. H. Holdsworth writing in 1841 repeats a precisely similar legend about Thomas Newcomen which was once current in Dartmouth. The notion that in the passage of thousands of years no man before Newcomen or Watt had possessed enough mother wit to observe this simple phenomenon and speculate upon it is fantastic in its absurdity. The power of steam had been known to man for unnumbered centuries; what was lacking was the intellectual concentration, the sheer dogged persistence which was required before such a power could be tamed and harnessed in the service of man.

The pursuit and application of knowledge in the pagan classical world was not inhibited by philosophical scruples but its direction was profoundly influenced by the way that world was organized. So long as the labour of slaves could be employed upon an immense scale to supply the needs of a fortunate minority that minority had no incentive to seek mechanical substitutes for human toil. The great Greek and Roman engineers such as Ctsebius and Vitruvius possessed a remarkable knowledge of the properties of steam and the expansive power of hot air but they made no attempt to apply their knowledge to useful purpose. The Whirling Aeolipyle or 'Ball of Aeolus' described in the *Pneumatica* of Heron of Alexandria (AD 100) employs, in its simplest form, the principle of the reaction turbine, yet, like many another ingenious gadget it made no claim to be other than a toy 'to keep a drowsy emperor awake'.

Very different considerations limited the pursuit of scientific knowledge in medieval Christendom. Although the men of the Middle Ages widely employed the elements of wind and water to turn their mills, they considered it both impious and dangerous to tamper with nature's more potent forces. The single-minded pursuit of knowledge for its own sake was a perilous undertaking. The men of the Renaissance dismissed this philosophy as so much superstition and the great forward march of technology began.

In Newcomen's day there were still some who had doubts but in the second half of the eighteenth century, any doubts of this kind were swept away on the tide of triumphant scientific conquest.

It is significant that the first attempts to harness the power of steam in Renaissance Europe followed the classical pattern very closely in that they were inspired by the desire to edify or amuse

1 Heron's aeolipyle, about AD 100, from British Museum, Burney MS 81, sixteenth century. This ingenious gadget demonstrated the principle of the reaction turbine.

and not by any serious utilitarian motive. Such were the devices — no more than laboratory experiments — produced by Giambattista della Porta (1536-1605) in Naples and Salomon de Caus (1576-1626) in England. Both simply used the pressure of steam generated in a closed vessel or boiler as a means of forcing water up a pipe. In della Porta's apparatus steam was passed from the boiler into the upper part of a closed tank, thus driving out through an ascending pipe the water in the lower portion. De Caus's arrangement was even simpler. It consisted merely of a small spherical copper boiler from which water could be expelled through a vertical pipe. The pipe extended almost to the bottom of the sphere and its outlet was controlled by a stop-cock.

This de Caus apparatus was almost precisely the same as the bronze ball or simple aeolipyle of Ctsebius which (according to Vitruvius), was used throughout the Roman world to provide a blast for fires and small furnaces, but whether de Caus re-invented it or was a student of Vitruvius we do not know. Of French parentage, de Caus was a landscape gardener who became an adept in the design and construction of waterwheels, pumps and other devices for working the fountains which he installed for his wealthy clients. After working in France and Italy he came to England in 1609 or 1610 and in 1611 he was planning a waterwheel-driven pumping installation for Hatfield House. He next worked at Greenwich Palace and was then engaged by Prince Henry, eldest son of James I, to ornament the gardens of his house at Richmond. In 1612 he published in London a work on perspective in which, on the strength of this royal patronage, he proudly styled himself 'Ingenieur du Serenissime Prince de Galles'.

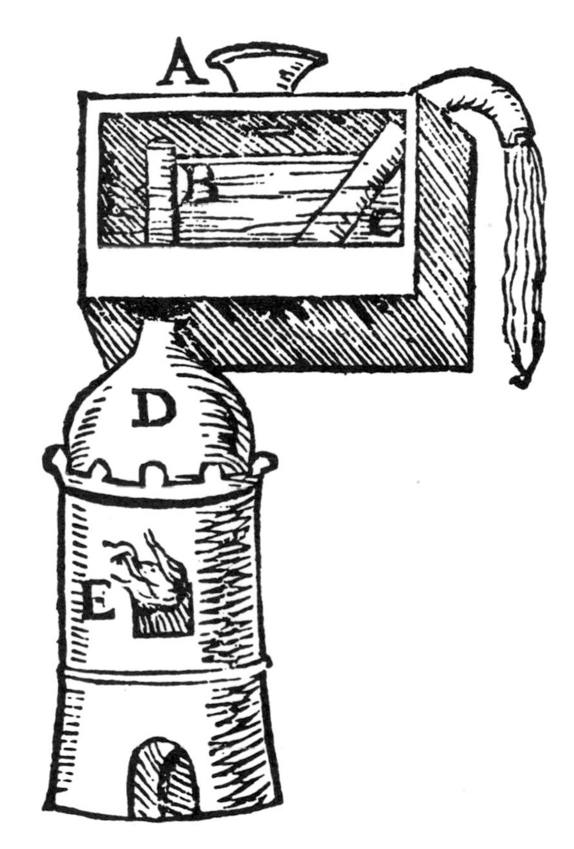

2 Della Porta's steam-pressure apparatus. From his *Spiritali*, 1606. The steam generated in a boiler forced water up a pipe.

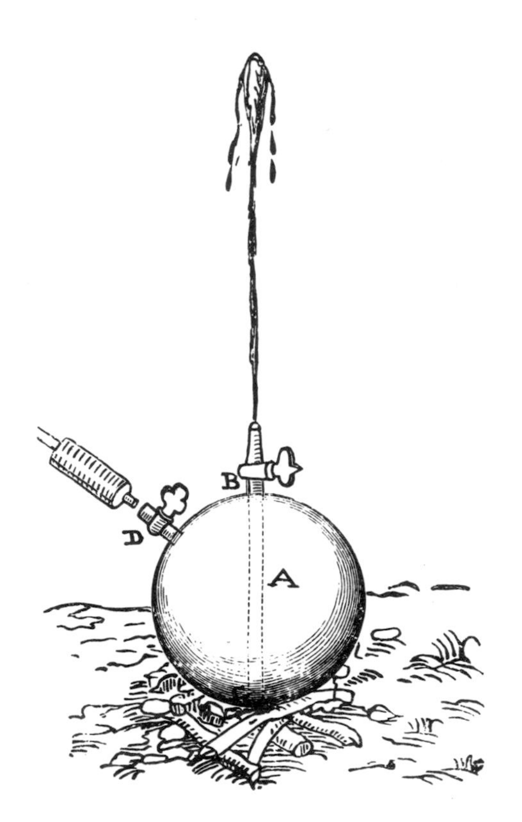

3 De Caus's steam pressure fountain, about 1611. A small copper boiler from which water could be expelled through a vertical pipe by steam pressure.

The young prince typified the new inquiring spirit that was abroad and took a keen interest in the ideas of de Caus, but unfortunately he died in 1612. A few months later his sister Elizabeth married the Elector Palatine Frederick V and in 1613 de Caus left England in their retinue for Heidelberg where the Elector appointed him as his engineer and architect. It was in Heidelberg that de Caus produced his most celebrated book. It was first printed in Frankfurt in 1615, but a French translation was subsequently published in Paris under the title: '*Les Raisons des Forces mouvantes, avec diverses Machines tant utiles que plaisantes,* par Salomon de Caus, Ingenieur et Architecte du Roy'. The first part of this book is a kind of encyclopaedia of useful arts and it is in this part that the author illustrates and briefly describes his little device for raising water by the power of steam. The second part, dedicated to Princess Elizabeth, consists of a series of designs for grottoes and fountains, but in it de Caus explains that the designs and devices in his book were contrived in former days for Prince Henry, some to adorn his palace at Richmond and others, as he puts it, 'Pour satisfaire a sa gentille curiosité, qui desiroit tousjours voir et cognoistre quelque chose de nouveau' — an explanation which expresses the moving spirit of the age very aptly.

Rhys Jenkins inferred from this that the little steam water-lifter was probably one of the devices which de Caus contrived in his Richmond days to satisfy the curiosity of a prince and that in this way he influenced two men who were about the Court of St James at that time. If this inference is correct, and it is a very plausible one, brief though de Caus's visit to England was, he sowed a seed which germinated slowly in other minds until it ultimately grew into a mighty tree. One of these two men was a Scot named David Ramsay who had come to London on the accession of James I and who was appointed Groom of the Bedchamber to Prince Henry. The other was young Edward Somerset whose grandfather was Keeper of the Privy Seal and who was to become the second Marquis of Worcester (1601-67).

A significant outcome of the new ferment of ideas was the Statute of Monopolies which was passed in 1624. This Act consolidated the law governing the principles under which letters patent were granted for new inventions. David Ramsey, either singly or conjointly with others, was one of the most prolific of early patentees. Many of his ideas are not relevant in this context, but among those covered by his patent No 50 of 1631 are the three following:

To Raise Water from Lowe Pitts by Fire.

To Make any Sort of Mills to goe on Standing Waters by Continual Moc'on without the Helpe of Windes, Waite or Horse.

To make Boates, Shippes and Barges to goe against strong Winde and Tyde.

Whatever ideas Ramsay may have had on these subjects they are not explained in the patent and there is no evidence that any of them ever achieved a practical shape. The significance of his patent is that it indicates the direction in which men's minds were turning. Unlike de Caus, Ramsay was not concerned to make new powers play tricks for the amusement of princelings but to harness them to utilitarian purpose. Henceforward priority of human need would begin to replace the whim of the aristocratic patron as the arbiter of inventive effort. Ramsay's list of inventions reflects, either by chance or design, the order of priorities as it was in his day. The propulsion of ships or mills by powers other than those of the wind or water may have become an intriguing subject for speculation, but it was not yet a pressing need. The old powers were still adequate to meet the trade demands of the time and it would be many years before Ramsay's cloudy concepts would materialize as powerful intricate creatures of iron and brass. With the drainage of 'Lowe Pitts' it was quite otherwise. Shaft mining began in Devon and Cornwall by 1500 and men had first been confronted by this problem by the time of the first Queen Elizabeth and with every year that passed, as shallow and easier levels were worked out and miners were forced to delve deeper in their quest for coal and mineral ores, the need to find a solution became more acute. It was to the drainage of mines, therefore, that inventive effort was first directed and it was from the ultimate solution to this problem that other power applications would stem.

The mantle of David Ramsey fell upon his junior, Edward Somerset, Marquis of Worcester who in 1663 secured an Act of Parliament which entitled him to: 'Receive the Benefit and Profit of a Water-Commanding Engine by him invented' for a term of 99 years. Newcomen's contemporary, Henry Beighton, hailed the Marquis in a long poem (published in 1725 and quoted at the beginning of this book) as the originator of the machine for raising water by the force of fire, but whether he merits such a distinction is extremely doubtful. The most that can be said for him is that he evidently did construct a machine upon a much larger scale than de Caus's apparatus or the equally toy-like steam impelled wheel described by the Italian Giovanni Branca in 1629. The machine was made at a cannon foundry at Vauxhall, the lease of which the Marquis purchased from the Crown on behalf of his collaborator, Kaspar Kaltoff. It is absolutely certain that this device was not a practical success and failed to fulfil the extravagant claims of its proud inventor. Robert Stuart's *Descriptive History of the Steam Engine* features two plates which purport to show the Marquis's engine, but these are purely conjectural. Two visitors to England in 1663 and 1669, Samuel Sorbière, historian to the King of France and Cosimo III de' Medici, Grand Duke of Tuscany, both refer to the 'Hydraulic Machine' of Worcester's invention which they saw at Vauxhall and which could raise water 40ft by the power of one man. But they do not mention fire or the power of steam and W. H. Thorpe, in a paper on the subject, concludes that what these travellers saw was simply a manual pump. Stuart's illustrations were based on Worcester's own writings and the information they give is very scanty. Although Worcester promised to write a full account of his inventions, he never did so and all he gave the world were two short pamphlets whose deliberate aim seems to have been to mystify and not to instruct. Their high-flown language reveals the Marquis as a showman rather than an inventor. Modesty was evidently not his strong suit.

The first of these pamphlets was written in 1659 and published in 1663 under the title: *A Century of the Names and Scantlings of the Marquis of Worcester's Inventions.* It is a passage in this which takes us as near as we are ever allowed to get to the great invention. It reads as follows:

> I have taken a piece of whole Cannon, whereof the end was burst, and filled it three-quarters full of water, stopping and securing up the broken end, as also the Touch-hole, and making a constant fire under it; within 24 hours it burst, and made a great crack; So that having a way to make my Vessels, so that they are strengthened by the force within them, and the one to fill after the other. I have seen the water run like a constant Fountaine-stream forty foot high; one Vessel of Water rarefied by fire driveth up forty of cold water. And a man that tends the work is but to turn two Cocks, that one Vessel of Water being consumed, another begins to force and refill with cold water, and so successively, the fire being tended and kept constant, which the self-same Person may likewise abundantly perform in the interim between the necessity of turning the said Cocks.

It has been assumed from this description that the machine consisted of a boiler connected to one or two receivers from which water could be forced by the admission of steam through control cocks. If this surmise is correct then Worcester's machine was an enlarged and refined version of della Porta's apparatus, the substitution of two receivers for the latter's closed tank ensuring continuity in operation.

Next in succession to the Marquis of Worcester was Sir Samuel Morland (1625-95), 'Master of Mechanicks' to Charles II. He served the King while in exile and after the Restoration was rewarded for his services with a baronetcy. In 1675 Morland purchased the Vauxhall Foundry property which, since the death of the Marquis in 1667, had become a sugar bakery. He is said to have filled the house and grounds with ingenious contrivances. Whereas the volatile Marquis of Worcester dissipated his inventive talents upon such oddities as 'speaking statues' and carriages from which the horses could be instantly released if they became too restive, Morland was an inventor of more sober merit. His most notable achievement was the packed-plunger force pump. For the orthodox suction pump in which the bucket or piston must necessarily be a close fit in the cylinder — a requirement not easily satisfied in the seventeenth-century — Morland substituted a solid cylindrical plunger and closed the top of the pump barrel with a leather gland through which the plunger entered it. With such an arrangement it was not necessary to make the plunger a snug fit in the barrel.

In 1682, State Papers recorded that: 'Sir Samuel Morland has lately shown the King a plain proof of two several and distinct trials of a new invention for raising any quantity of water to any height by the help of fire alone.' Charles was so impressed by what he had seen that by his order a warrant for the grant of a patent to Morland was drawn up in December of the same year and in this document the King stated that: 'we are fully satisfied [that the invention] is altogether new and may be of great use for the clering of all sortes of mines, and also applicable to divers kinds of manufactures within our dominions.'

The design of this engine which so impressed Charles is not known. Morland's only reference to the subject is a chapter in a small book which he wrote in French while he was in France in 1683. The book is entitled *Elevation des Eaux, par toute sorte de Machines* and in translation the relevant chapter heading reads: 'The Principles of the New Force of Fire, invented by the Chevalier Morland, in the Year 1682, and presented to his Christian Majesty, 1683.' 'Principles' is here the operative word for the following text gives no hint whatever of the nature of Morland's machine. It does, however, include a table giving the weights which may be raised by applying the force of steam in cylinders of differing lengths and diameters up to 10ft and 6ft respectively.

Despite the evidence of State Papers it was seriously doubted by subsequent writers whether Morland ever constructed an engine at all. It is only comparatively recently (1936) that H. W. Dickinson drew attention to a contemporary sketch which may represent his machine. This occurs in the diary of Roger North (1653-1734), a son of Sir Dudley North and a barrister of the Middle Temple. He was Solicitor-General to the Duke of York who, when he succeeded to the throne, appointed North Attorney-General to the Queen. His diary sketch shows a boiler the steam space of which is connected by a 'Y' pipe and cocks to the bottoms of two vertical cylinders or 'sockets' as North calls them. From the tops of these cylinders emerge two plungers, one extended and marked 'plugg blown up' the other retracted and marked 'plugg sinking down'. Considering that North was not a technician his accompanying description is remarkably lucid. It reads as follows:

> The Rising of ye pluggs are ordered to turne a wheel by a toothed barr [a marginal sketch shows a toothed rack and pinion], wch when at the top, is struck loos by a catch or snack and then ye barr falls downe, & with its weight turnes ye wheelwork, wch shutts out ye steam from that pipe or socket, by a stop cock, & pari passu opens the other and then that riseth in like manner, and so they play alternately without help.
>
> The uses are derived from the wheel wch these rising barrs work upon, thus alternately as ye occasion is, ffor if a Motion be given either to and fro or continuall, it may be applyed by wheel work to almost all occasions, But this I saw onely in model.

This is the earliest known employment of steam from a separate boiler to act upon a movable piston or plunger in a closed cylinder. It is also the

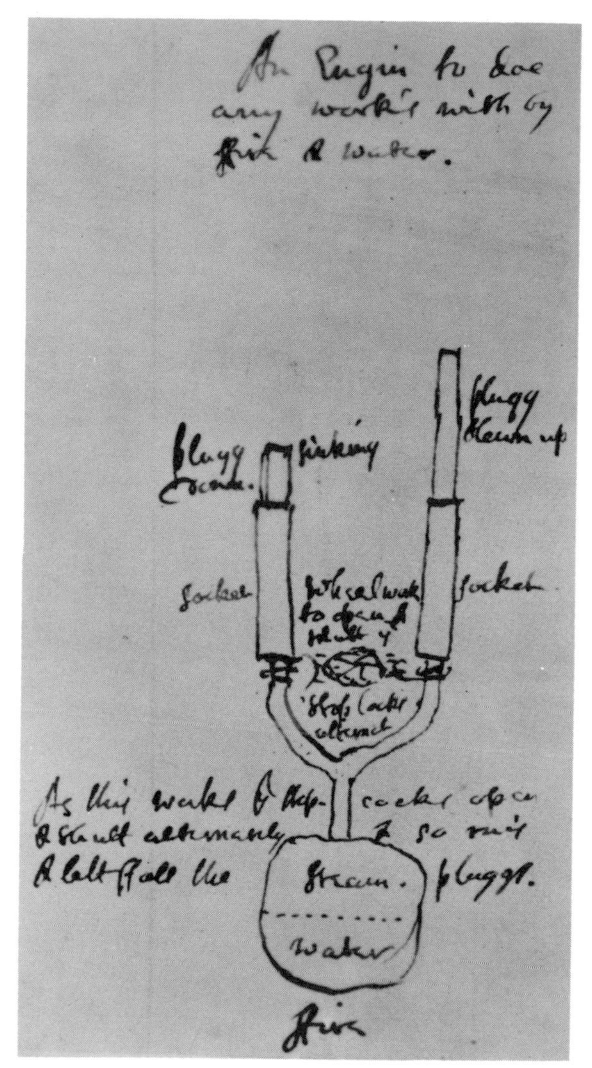

4 Sketch of an engine, believed to be Morland's, by Roger North, about 1680. 'An engin to doe any work's with by fire and water. As this works ye stop-cocks open & shutt alternately and so raise & lett fall the pluggs.'

oldest known example of automatic or self-acting valve gear, the vertical falling bar used to turn the steam cocks anticipating the plug rod used on the beam engine. It is all the more unfortunate, therefore, that North gives no clue as to where he saw this ingenious machine or by whom it was made. Moreover, the sketch gives no indication as to how each plunger made its down stroke after the steam which had blown it up had been shut off. However, we must remember that North was an amateur recording from recollection.

Rhys Jenkins gives good grounds for the theory — it can be no more — that the engine North describes was Morland's. North and Morland were well acquainted with each other through the agency of North's brother, Francis North, Lord Guilford, Lord Chancellor and Keeper of the Great Seal. After the death of Lord Guilford in 1685, Roger North wrote a life of his brother in which he refers to the fact that Guilford and Morland were 'good philosophical friends' and describes the ingenious contrivances which he saw on one occasion when he and his brother visited Morland's house. There is also the evidence of the engine itself with its use of plungers so similar to those employed by Morland in his pumps. North's concluding statement: 'But this I saw onely in model' may refer, not to the complete machine but to its adaptation to rotative motion and the refinement of the self-acting valve gear. His sketch shows neither of these features and it is possible that the engine which so impressed Charles II was a larger but simpler reciprocating pump with manually-operated steam valves.

Whoever was responsible for the machine which Roger North portrayed richly deserves a place in any history of the steam engine despite the fact that it appears to have been a premature 'sport', a sterile *tour de force* which had no influence upon later developments. If the engine was Morland's, the reason for its failure to bear fruit may lie in its inventor's tragic history. Charles II sent his 'Master of Mechanicks' to France to assist and advise Louis XIV on the problem of providing a water supply to his new palace at Versailles which was greatly troubling him. £2,500 in English money were spent on the works proposed by Morland, but they were not a success and the engineer fell into disgrace. He also became so short of money that for some time he was unable to afford his passage home. When he finally managed to return in 1685, sixty years old, impoverished and dispirited, he soon suffered a fresh blow on the death of his friend and patron Charles II. The rest of his life was spent in comparative obscurity at his home in Hammersmith. When John Evelyn the diarist visited him there in 1695, the year before his death, he found Morland aged, deaf and blind but still indomitable. To overcome his disabilities he had invented a speaking trumpet and a special calendar readable by touch.

The path of progress that passed Sir Samuel Morland by leads directly from the Marquis of Worcester to Thomas Savery. There is a marked resemblance between Stuart's conjectural representation of Worcester's machine and Savery's engine but in their working principles there was a vital difference. For whereas the Marquis proposed using steam pressure alone, Savery also called to his aid another power, atmospheric pressure. So far this chapter has been concerned only with attempts to harness steam power and no mention has been made of man's efforts to understand and command the phenomenon of atmospheric pressure. It is necessary now to go back again through time in order to trace this second converging line of development from its first beginnings down to the date when the two principles were wedded in Savery's machine.

The Discovery and Application of Atmospheric Pressure

While the power which steam can exert was one of the more obvious manifestations of natural forces, it was not until the Renaissance that men learnt to account correctly for the phenomena caused by the weight of the earth's atmosphere. This was one of nature's secrets with which none had dared to tamper. Nature abhorred a vacuum, a nothingness, for, so the medieval schoolmen taught, the whole universe was filled with matter of some kind. Such was the divine order and the idea that man might exclude all such matter and so create a nothingness was unthinkable in its impiety. Only God could with impunity so interfere with the order of His creation.

So matters stood until 1641 when the engineers employed by Cosimo II de' Medici, Grand Duke of Tuscany, constructed a suction pump to draw water from a 50ft well at the Duke's villa in Florence. They were completely mystified by the fact that their utmost efforts with the pump would not draw the water more than 32ft up the suction pipe, whereas according to the accepted law of the *horror vacui* there should be no limit to the height to which the water would follow the pump bucket. They took their problem to Galileo (1564-1642), but that great philosopher was just as puzzled by the phenomenon as they were. He undertook to investigate it, but for the present he could only reply lamely that there were evidently limits to nature's abhorrence.

Galileo died in the following year before he had reached any conclusion, but his investigations were carried on by his favourite pupil and amanuensis Evangelista Torricelli of Faenza (1608-47). It was Torricelli who first deduced that the pump phenomenon was caused by the fact that the earth's atmosphere had weight. The column of water *in vacuo* in the pump suction pipe would, he argued, only rise until its weight equalled that of the atmosphere; there could then be no further rise since a state of equilibrium had been reached. He further argued that if this theory was correct the maximum height which a column of a heavier liquid would attain in vacuum would be reduced in direct proportion to its increase in weight. He therefore repeated the experiment with mercury, which is about fourteen times heavier than water, and found that the result confirmed his theory. The column rose only 29in before equilibrium was reached. In this classic experiment, which Torricelli carried out with the assistance of his own pupil Vincenzio Viviani (1621-1703) in 1643, he not only proved that the atmosphere had weight but invented the barometer.

The brilliant French savant Blaise Pascal (1623-62) carried the proof of Torricelli's proposition a stage further. If the earth's atmosphere had weight, Pascal argued, then the pressure it exerts should be greater at sea level than at high altitudes just as the pressure of water increases with its depth. When Pascal advanced his theory in 1647 he was living at Rouen where he could not put it to the test. He therefore persuaded his brother-in-law, Florin Périer, who lived in the mountains of the Auvergne near Pascal's birthplace at Clermont, to make the experiment. Assisted by a party of friends, Périer ascended the 4,800ft peak of the Puy de Dôme on 26 September 1648, carrying with him a mercurial barometer tube 4ft long. The height of the mercury had first been checked and found to

5 Von Guericke's experiment with vacuum about 1672. The rope was attached to a piston in a cylinder and twenty men ordered to haul up the piston. An exhausted sphere was then connected and atmospheric pressure drove the piston down, overcoming the efforts of the men to restrain it.

correspond exactly with that in a second similar tube which was left at the base of the mountain. On the summit it was found that the mercury column had fallen by $3\frac{1}{2}$in. A series of readings taken on the descent showed that the mercury was steadily rising until, on regaining their starting point, it was found that the levels of the two instruments again corresponded.

By proving that the atmosphere had weight and could therefore exert pressure, man had not only found the answer to phenomena which had puzzled him but he had also discovered a potential source of power, provided he could perfect some ready means by which he could create a vacuum.

Progress in this direction was initiated by Otto von Guericke (1602-86) of Magdeburg. Although they were contemporaries, Guericke was unaware of the work of Torricelli and Pascal, but his interest in astronomy led him to speculate whether or not there was air in outer space. He believed that there was not, because he reasoned that the friction of an atmosphere would slow down the stars and planets in their courses. It was the notion that outer space might be a vacuum which led to his attempts to create such a vacuum. His idea was to fill a closed vessel with water and then to pump out the water, at the same time preventing the atmosphere from taking its place. After several abortive attempts with wooden casks, Guericke succeeded in this way in exhausting a copper sphere to such a degree that it collapsed inwards with a loud report. Greatly encouraged, he tried again with a stronger sphere and a more powerful pump. This time he was completely successful; the sphere was exhausted and when a stop-cock in its side was turned on the air rushed in with a force which surprised him.

So far Guericke had used water as a medium to exclude the air from a receiver, but now he devised an air pump with which he could exhaust the receiver directly. With this new pump he made two spectacular and historic experiments. For the first he constructed two identical hemispheres of copper, 12ft in diameter, which, when fitted together like the two halves of a child's Easter egg, could be sealed at their rims by an airtight leather joint. Before the Emperor Ferdinand III and members of his government Guericke demonstrated that when he had exhausted these united hemispheres the efforts of sixteen horses could not pull them apart.

In the light of the future, Guericke's second experiment was even more significant. He constructed a cylinder with a tight-fitting piston. The cylinder was securely anchored in a vertical position with its open end uppermost and a strong rope attached to the piston rod was passed over a pulley directly over it. Twenty men grasped each a strand of this rope and were ordered to haul the piston to the top of the cylinder and hold it there against the partial vacuum so created. Guericke then produced a copper sphere which he had previously exhausted with his air pump and connected it to the base of the cylinder via a pipe

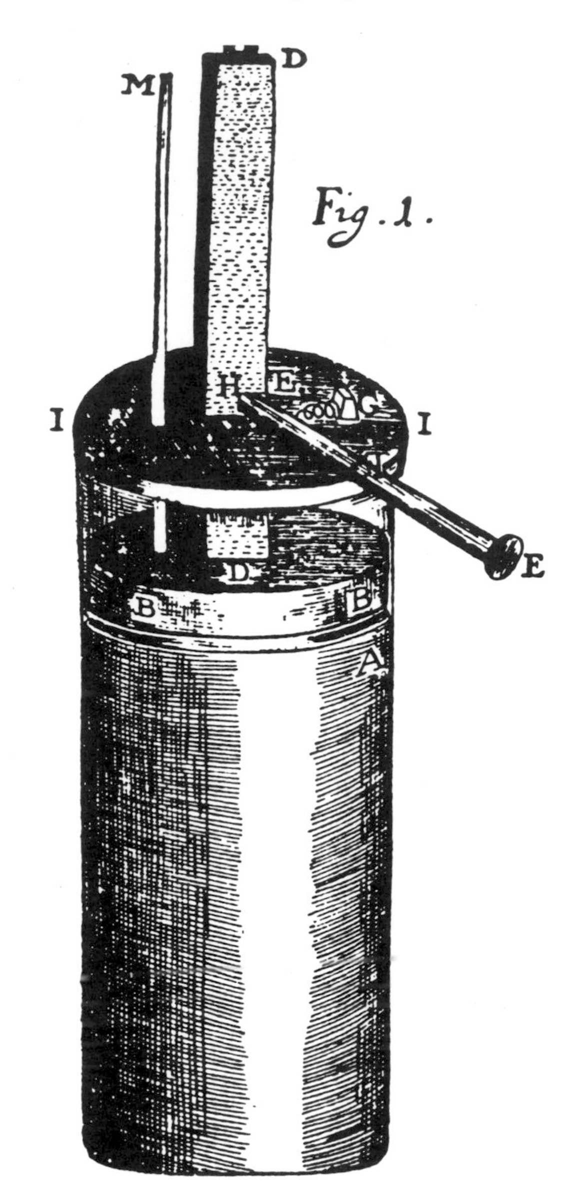

6 Papin's cylinder and piston vacuum apparatus, from *Acta Eruditorum*, 1690. Steam was raised in the cylinder A, under the piston B, which raised the piston rod D, where it was held by catch E at H. On the fire being removed, the steam condensed and the catch being released the piston descended, raising a weight by means of a cord and a pulley.

and two stop-cocks. As soon as the cocks were opened the residual air in the cylinder rushed into the sphere with the effect that atmospheric pressure drove the piston downwards with great force despite the efforts of the twenty astonished men to restrain it.

The man who combined the findings of Torricelli, Pascal and Guericke and brought these fruits of his Continental travels home to England was the Hon Robert Boyle (1627-91), seventh son of Richard, Earl of Cork and Orrery. Boyle was an indefatigable experimentalist who carried on his researches in London and Oxford despite the stresses and strains of civil war. After the war he became one of the founders of the Royal Society. With the assistance of Robert Hooke (1635-1703), the physicist and mathematician, he evolved an improved type of air pump. Boyle's pump is still preserved by the Royal Society and is the prototype of those used today. It was with its aid that Boyle formulated the law of gas pressures which is known by his name. He also used it to repeat and confirm the results of the experiment which Périer had carried out for Pascal on the Puy de Dôme but without doing any mountaineering.

Guericke's experiments had convincingly shown that the weight of the atmosphere was potentially a useful source of power, but its practical application depended upon the discovery of some means of creating a vacuum other than that of the air pump. Surprising though it may seem to us, men's thoughts first

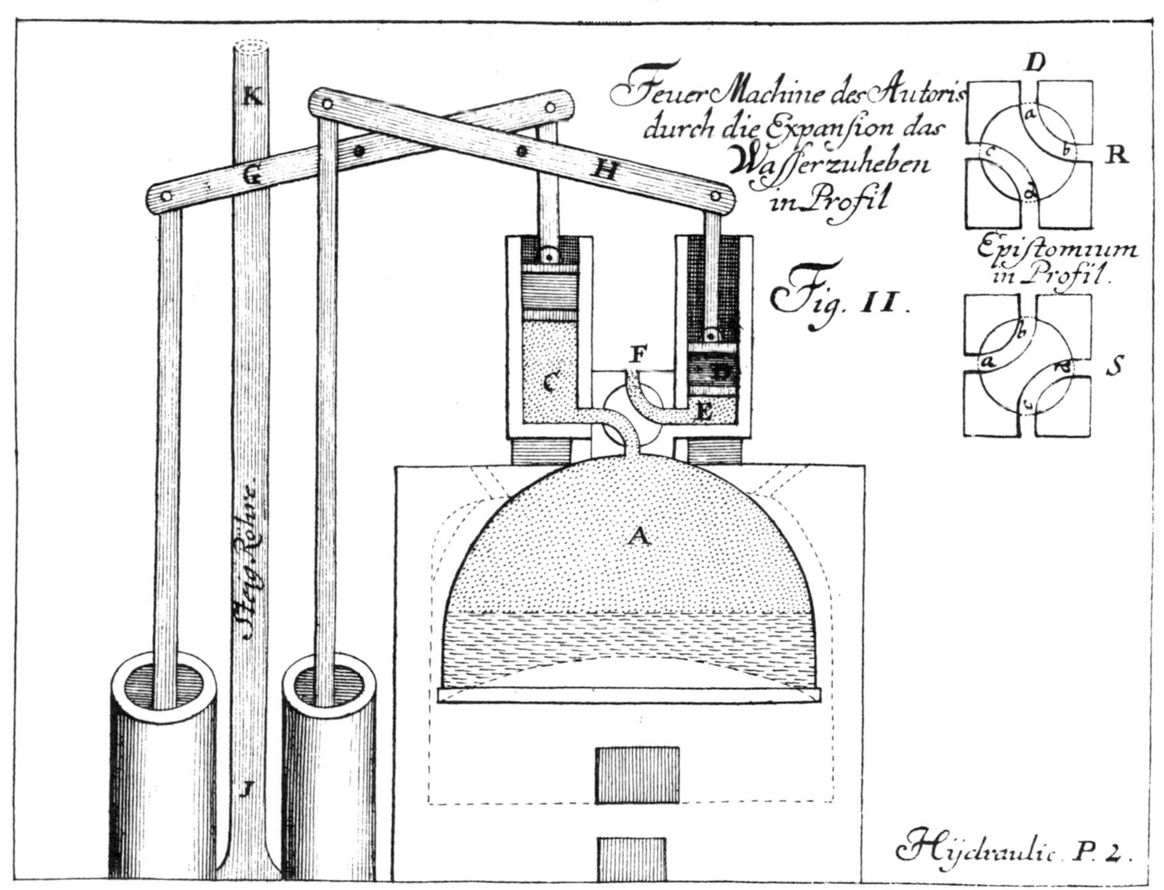

7 A proposed high-pressure engine by Leupold in which the steam is admitted by the valve into each cylinder alternately. Pistons then operate the pumps via lever beams to pump water up the output pipe KJ.

turned to gunpowder as a means of achieving this end, the idea being to utilize the explosive force of a small charge of powder to expel the air from a closed vessel. In December 1661 Sir Samuel Morland was granted a patent for an engine to raise water from mines 'by the force of Aire and Powder conjointly', but if such an apparatus was ever made nothing is known of it and the next steps were taken on the other side of the English Channel.

The problem of supplying the new Palace of Versailles with water, which proved the downfall of Morland, exercised other brains besides his. In 1678, Jean de Hautefeuille proposed using two cisterns into which water would be drawn by the vacuum created by a small charge of powder, the air being thus expelled from the cisterns through non-return valves.The two cisterns were to be exhausted alternately in this way to maintain continuity of supply. The great Dutch astronomer Christiaan Huygens (1629-95) and his assistant Denis Papin (1647-1712) began experimenting on similar lines and with the same end in view at the Académie Royale des Sciences, the equivalent of the British Royal Society, which was established in Paris by Louis XIV in 1666. Instead of a cistern, they used a cylinder and piston, one end of the cylinder being closed and the other open to atmosphere. When a small charge of powder had expelled the air from the closed end of the cylinder through non-return valves in the form of tubes of soft leather, atmospheric pressure drove the piston down.

In 1675, Denis Papin, who was a Huguenot, left France to escape religious persecution and came to London where he worked, first for Robert Boyle and later, through Hooke's introduction, for the Royal Society. Papin demonstrated before the society several ingenious ideas. One was his 'new Digester or Engine for softening Bones', which was the world's first pressure cooker, and another was his double-acting air pump. In connection with the latter Papin proposed using a twin-cylinder air pump driven by a waterwheel to operate at a distance by pipeline a cylinder and piston-type atmospheric engine. This was the first proposal for power *transmission* in history but it did not contribute anything towards the problem of power generation. It had become obvious to Papin and his contemporaries that the use of gunpowder could never be a satisfactory solution and it was Papin who hit upon the far more practical alternative of creating a vacuum by the condensation of steam. He constructed a cylinder and piston apparatus in which the gunpowder charge in the closed end of the cylinder was replaced by a small quantity of water. He then lit a fire under the cylinder until the water boiled. The piston and the rod attached to it rose to the top of the cylinder where it was held by a catch or detent and the cylinder filled with steam, the air being driven out through a non-return valve. The fire being then taken away, the steam condensed as the cylinder cooled, a vacuum was created and when Papin released the detent the piston was driven down the cylinder by atmospheric pressure, raising a considerable weight by means of a cord and pulley.

Denis Papin, at some time between 1690 and 1695, thus discovered and demonstrated the principle of the atmospheric engine which Thomas Newcomen would later apply to such great effect. Yet, tantalizingly, having reached the very brink of practical success, the brilliant Papin turned aside. His engine was only a laboratory toy and, arguing that its reproduction on a large scale involved insuperable practical difficulties, he devoted himself to unsuccessful attempts to harness steam pressure without using the mechanical medium of cylinder and piston.

Thomas Savery and His Pump

We have now reached the point in this story of evolution where the two sources of power, the expansive force of steam under pressure and the weight of the earth's atmosphere, were combined for the first time in Thomas Savery's steam pump. Moreover, Savery's invention was no scientific toy like its predecessors but a practical working machine although, as we shall see, its inherent defects were such that its value in application was extremely limited.

Like Newcomen, Thomas Savery was a Devonian. The Saverys were a family of prosperous merchants in Totnes who, about 1614, acquired the nearby manors of Shilston and Spriddlescombe in the parish of Modbury. The local registers reveal no record of Thomas Savery's birth, but he was reputedly born at Shilston about 1650. In some ways he is as elusive a figure as Newcomen. There is no certain portrait of him and his early career is shrouded in mystery, but there is what is claimed as his portrait 'Engraved by W. T. Fry from an original painting' in the 1827 reprint of *The Miner's Friend.* After 1700 he is referred to as Captain Savery, but there is no evidence to show how he came by this title. His name does not feature in the Army Lists for the period, but his translation of a work on fortification and other shreds of evidence suggest that he may have worked as a military engineer. Another theory is that he was given the title in Cornwall where the miners customarily referred to engineers and mine-managers as captains. Yet other writers refer to Savery as a sea captain, but this is highly improbable, as in a passage of his writing he disclaimed any intimate knowledge of ships and the sea.

Savery's family were well known in the West Country and he may have been a merchant at Exeter for a period. In 1673 Charles II issued a writ requiring the mayor and aldermen of the town to cease from molesting one Thomas Savery, a freeman of the Merchant Adventurers Company who had 'received many losses and particularly in the last Dutch Wars'.

At the beginning of the War of the Spanish Succession in 1702, the Admiralty set up a special commission for the sick and wounded and to make arrangements for the custody and exchange of prisoners of war. This commission had agents not only at the main overseas stations but at every important seaport in the country. The commissioners, including leading physicians and apothecaries, met three times a week and their peak turnover was about £100,000 a year. The post of treasurer was therefore a most important one and it fell vacant in 1705 on the death of Richard Povey.

After a full investigation of his circumstances and the filing of suitable guarantees Prince George issued the commission to Savery for which he was to receive a salary of £200 per year, and had the use of an apartment. In 1705 Savery joined the Royal Society and became active in its affairs.

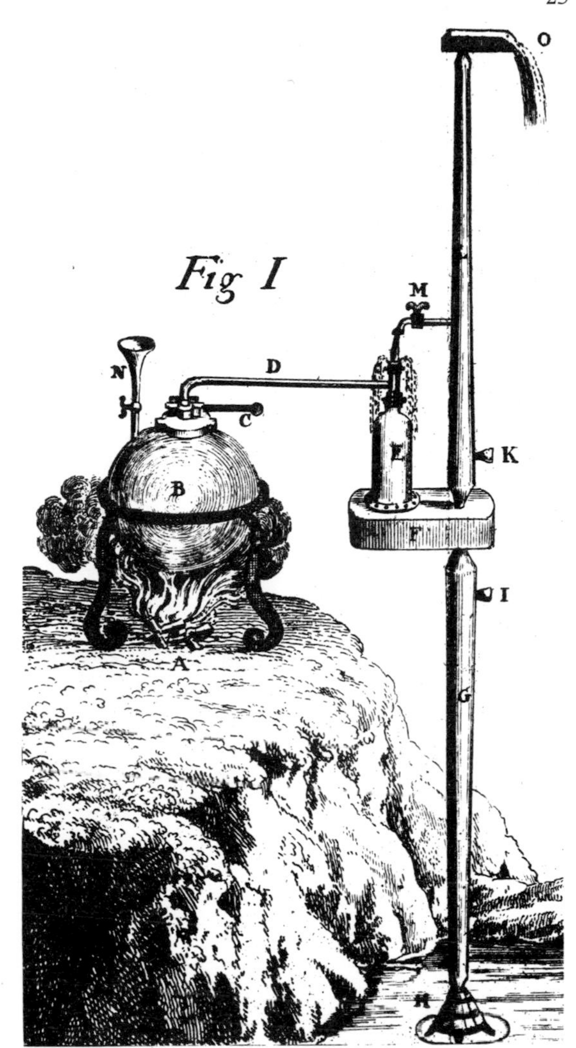

8 An early Savery pump with single boiler and receiver, about 1698. Steam under pressure in boiler B forces water from vessel E, passing non-return valve K to outflow. The steam in vessel E was condensed and raised water by vacuum up pipe G passing non-return valve I.

Due to the need for economies at the end of the war, Savery was relieved of his duties at the commission, with effect from 24 June 1713. For some time after he left his position, the treasury and the commissioners communicated with Savery in connection with payments and

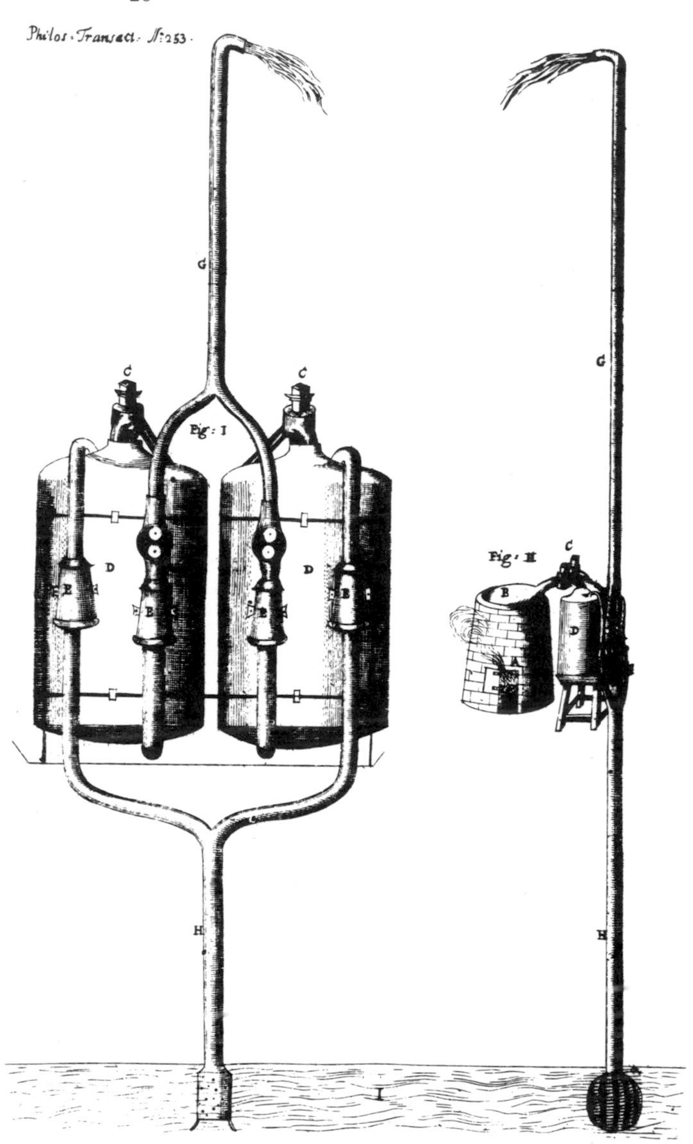

9 Savery's later design of pump with twin boilers and receivers, from *Philosophical Transactions.* By operation of the valves continuous flow could be achieved.

statements. His affairs were probably never settled for matters had not been cleared at his death in 1715 and administration of his estate was not granted until 1796 some 37 years after the death of his wife, Martha Savery, at the age of 104 in 1759.

Savery was the most prolific inventor of his day and on 25 July 1698 was granted the historic patent for 'Raising water by the impellent force of fire' which was to play an important part in the story of Thomas Newcomen. This original grant was for fourteen years, but an Act of Parliament in 1699 extended this term by twenty-one years, that is to say until 1733. In 1701 Scotland was included for the same term of years.

The principle of Savery's pump, which was the subject of this patent, was as follows. Steam was passed from a boiler into a closed receiver filled with water where its pressure forced the water through a non-return valve and up an ascending delivery pipe. When the rattling of the non-return or clack valve indicated that all the water had been expelled, the steam supply from the boiler was shut off and a cock on a branch pipe from the main delivery was turned on, releasing cold water which poured over the outside walls of the receiver, cooling it and so condensing the steam within. A vacuum was thus created in the receiver with the effect that water was forced up a 'suction' pipe and through a second non-return valve into the receiver by atmospheric pressure. When the receiver had been refilled in this way the cooling water was shut off, steam turned on and the cycle began again.

It seems evident that the first working model constructed by Savery consisted of two main units only, a single boiler and a single receiver, but the rapid development of the invention thereafter gives us the measure of Savery's ingenuity. In June 1699 he demonstrated to the Royal Society a pump with two receivers, each with its separate, hand-controlled steam supply. This ensured greater continuity of operation, one receiver being in vacuum while the other was under steam pressure. Roger North, in the diary which was mentioned earlier, describes a similar Savery pump in which the two steam admission cocks to the receivers were connected by 'wheelwork' in such a way that when one was opened the other was closed. This improvement must have been made either in the second half of 1699 or in 1700, for in 1702 Savery published a book, *The Miner's Friend,* written in 1701, in which he described and illustrated his pump in its final form. In this he has replaced the two interconnected steam cocks by a single valve, manually operated by a long lever. It consisted of a fan-shaped sector plate which covered and uncovered the two steam ports alternately as it was moved to and fro on its vertical axis. This was a development of great significance, for a similar type of sector valve, mechanically actuated, would later be used on Newcomen's engines. Moreover, on his earliest engines James Watt used a similar sector valve with the difference that, instead of a flat plate, it was made in the form of an inverted box, the steam exhausting from the cylinder to the separate condenser through this box. Although Watt later abandoned this type of sector valve in favour of separate drop valves for steam admission and exhaust, the evolutionary link between it and the slide valve is obvious. It could therefore be said that the seed of the idea of the slide valve of modern times was sown by Thomas Savery.

Savery made another important improvement on his pump of 1701. He had previously equipped his boiler with water level try-cocks, but on his earlier designs when these cocks showed that the water was almost exhausted it was necessary to draw the fire and shut down the pump in order to refill the boiler. Now, Savery refers to his steam generator as 'the great boiler' because he had overcome this disadvantage by connecting it to a second, smaller boiler which was, in fact, a water supplier and the world's first feed-water heater. This functioned in the following way. The second boiler was entirely filled with water from the pump's delivery pipe. When the attendant ascertained from the try-cocks that the water in the 'great boiler' was getting low he lit a fire under the second boiler until the pressure within it rose sufficiently to overcome a clack valve and so force its contents into the 'great boiler'. The fire under the second boiler was then either drawn or damped down and it was refilled. This was a tedious process and not without its hazards, but it did ensure that the pump could be kept at work continuously.

To Savery must go the credit for taking the first momentous step out of the scientist's laboratory and into the workshop. He asked a great deal of the craftsmen of his day, but if we are to believe him he eventually succeeded after much difficulty in achieving a creditable standard of workmanship. The cocks and pipework were of brass and the receivers and the spherical boilers upon their

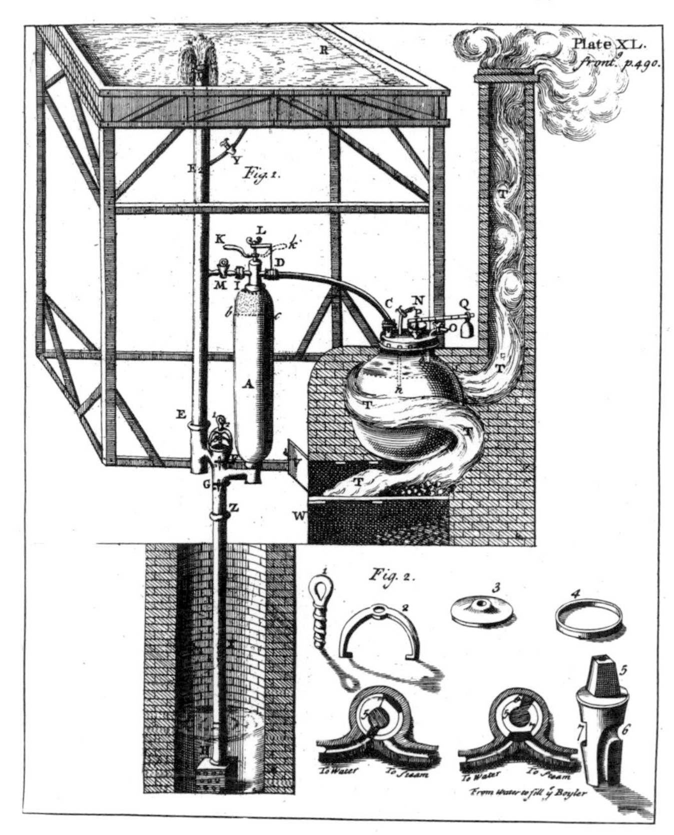

10 A Savery engine in use at a waterworks, raising water from a well to an elevated storage tank. From J. T. Desaguliers *Experimental Philosophy*, 1744.

firebrick furnaces were of beaten copper. The *Post Man* for 19-21 March 1702 carried the following announcement of Savery's activities:

> Captain Savery's Engines which raise Water by the force of Fire in any reasonable quantities and to any height being now brought to perfection and ready for publick use. These are to give notice to all Proprietors of Mines and Collieries which are incumbred with Water, that they may be furnished with Engines to drain the same, at his Workhouse in Salisbury Court, London, against the Old Playhouse, where it may be seen working on Wednesdays and Saturdays in every week from 3 to 6 in the afternoon, where they may be satisfied of the performance thereof, with less expense than any other force of Horse or Hands, and less subject to repair.

Salisbury Court, which extends from Fleet Street down to the river Thames, was thus the site of the world's first steam pump manufactory but, alas, this failed to fulfil the hopes of its ingenious founder. It seems that Savery abandoned the whole project in 1705 although according to some authorities he installed a few small single receiver pumps for private domestic water supply purposes after this date. The brutal fact was that for mine pumping, the most pressing need of the day, Savery's machine was quite unsuitable. The suction lift of the pump was strictly limited as the engineers of Cosimo de' Medici had found. Twenty feet was the greatest practicable length of suction pipe below the base of the receivers. The forcing lift was likewise limited to a figure not substantially exceeding this by the low steam pressure which was all that the boiler could carry. What happened when Savery tried to increase the forcing lift was told by Dr Desaguliers in his *Experimental Philosophy* (1744). He writes:

> I have known Captain Savery, at York Buildings make steam eight or ten times stronger than common air [ie 117 to 147 lb/in^2]; and then its Heat was so great that it would melt common soft Solder; and its Strength so great as to blow open several of the Joints of his Machine; so that he was forc'd to be at the Pains and Charge to have all his Joints solder'd with Spelter or hard Solder.

York Buildings was one of the waterworks which supplied London with water from the Thames and while we cannot fail to salute Savery for his courage it is scarcely necessary to add that his York Buildings pump was a failure. To have pumped water from deep mines by his method would have involved the installation of a series of

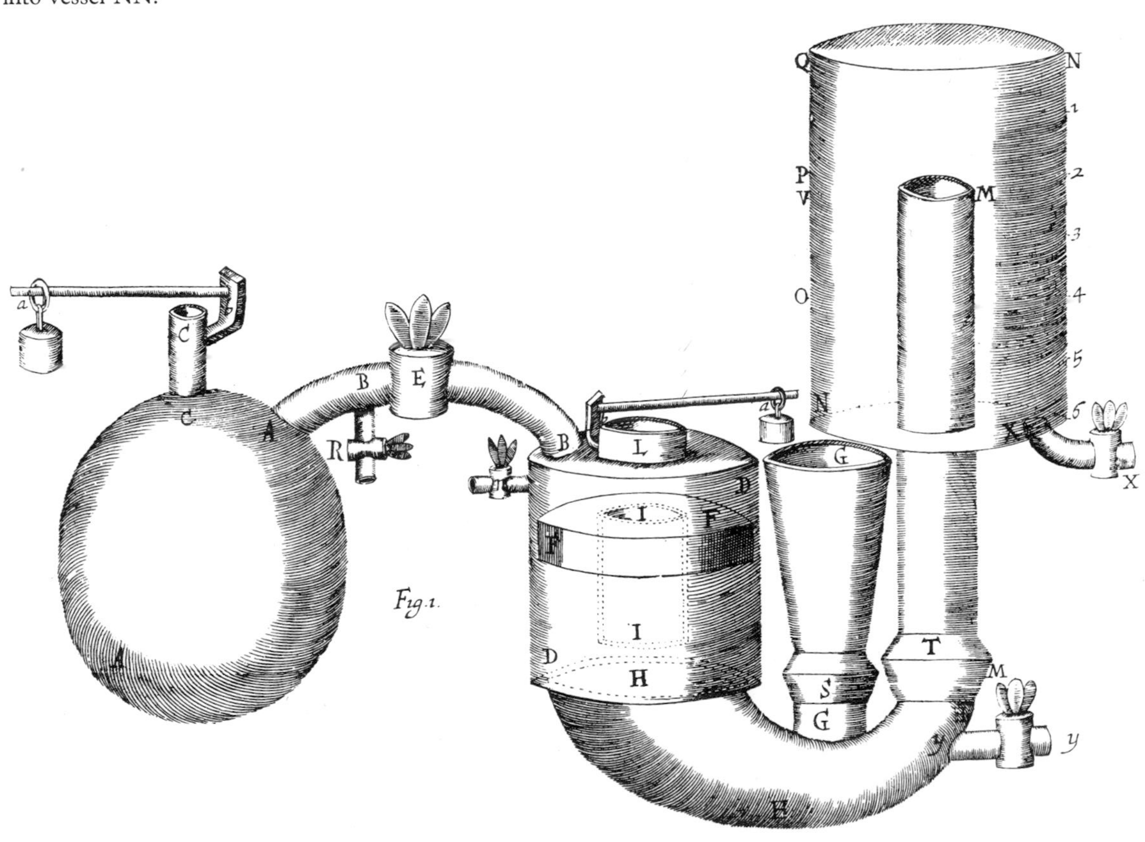

11 Denis Papin's improved Savery engine from a paper presented to the Royal Society in 1707. The piston FF, floating on the water, was intended to assist the steam from boiler AA to force water fed from GG into vessel NN.

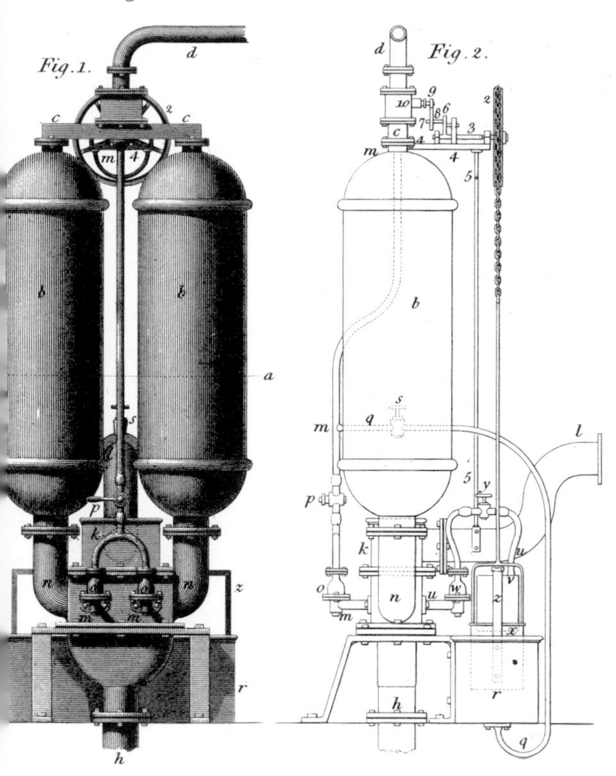

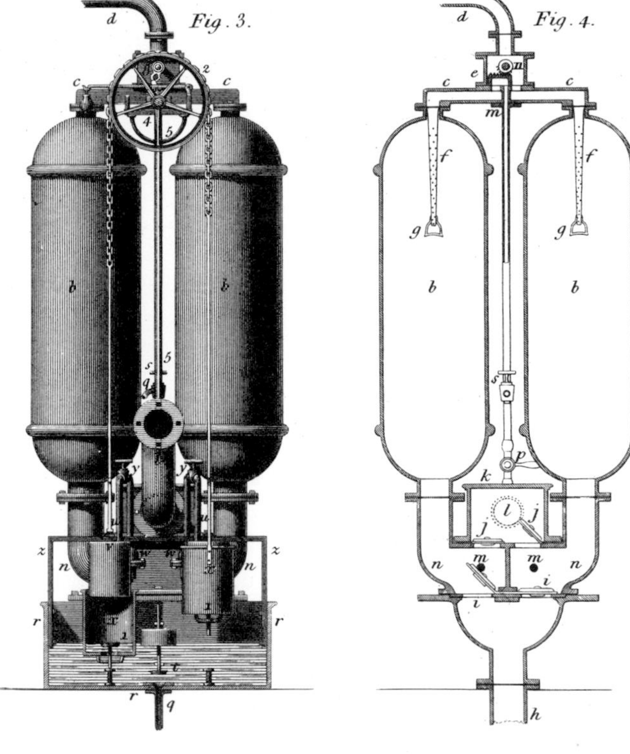

12 An improved Savery engine by J. Pontifex erected as late as about 1822 at the City Gas Works, London. Illustrated by Partington, *History of the Steam Engine.*

pumps below ground at vertical intervals of approximately 50ft, a system which was obviously impracticable, being at once too costly and too dangerous.

On the score of danger it is a curious fact that, despite the ingenuity he displayed in other directions, Savery never fitted his boilers with any form of safety valve. Credit for the invention of the weight safety valve goes to Denis Papin who first used it on his 'Digester' or pressure cooker. The safety factor of the engine was almost non-existent, though some of Savery's customers seem to have been blissfully unaware of this. 'How useful it is' wrote Switzer, 'in gardens and fountain works may or might have been seen in the garden of that right noble peer, the present Duke of Chandois, at his late house at Sion Hill, where the engine was placed under a delightful banquetting-house, and the water being forced up into a cistern on the top thereof, used to play a fountain contiguous thereto in a very delightful manner.' It is certain that the Duke's guests would have very willingly forgone the pleasure of watching the fountain play had they fully understood what was going on directly beneath their feet.

Curiously, despite the evident limitations of Savery's system, it continued to attract scientists and engineers long after the death of its inventor in May 1715 and long after Newcomen's engine had proved itself beyond dispute. The source of this attraction was that, unlike the Newcomen engine, Savery's pump required no heavy mechanical moving parts the friction of which, it was argued, was a source of great inefficiency. That during the forcing stroke of Savery's pump the steam from the boiler was brought into direct contact with the cold water in the receiver was, in fact, a much greater source of inefficiency than the frictional losses in Newcomen's engine. Savery's successors tried unsuccessfully to mitigate this defect by interposing between water and steam in the receiver various forms of 'floating pistons' or diaphragms. Desaguliers produced a variant of the Savery pump in which condensation in the receiver was brought about by internal water injection instead of by cooling the outside walls, and as late as 1776 William Blakey patented an unsuccessful engine of Savery type in which cold water and hot steam were divided by a cushion of air.

None of these attempts to overcome this basic fault of the Savery cycle were of any avail and the only practically successful Savery-type machines evaded the problem by using the suction part of the cycle only. The steam in the receiver having been condensed by cold water injection, the water thus drawn up into the receiver was discharged into a cistern where it was used to drive an overshot waterwheel. The steam and water injection valves were actuated from the waterwheel shaft and the wheel also drove a small force pump to supply the injection water. Joshua Rigley of Manchester built a number of these simple machines in Lancashire in the 1760s and John Farey describes and illustrates an engine of this type which was still used to drive lathes in a small works at St Pancras in the early years of the nineteenth century.

Despite the persistence of his successors, Savery's gallant effort must be judged a failure because it could not answer that most pressing question of his day: how to get rid of the water from deep mines. So acute was this problem that the mine owners would have tolerated its gross inefficiency provided the pump could do the job. Clearly it could not, for to do so it would require a steam pressure so high that contemporary technology could not contain it. So the line of development begun by Savery ends in the pulsometer pump and the future of the steam engine lay along a different path, the path that Guericke had first signposted and from which Denis Papin had so inexplicably turned aside. Until men had learned how to contain and control high pressure steam, the right way forward was to harness atmospheric pressure by means of cylinder and piston. The man who proved this to the world was Thomas Newcomen.

CHAPTER 2

Enter Thomas Newcomen

Unlike any of his predecessors, noblemen, 'philosophers', royal protégés, Thomas Newcomen was a practical tradesman, an ironmonger of Dartmouth. It was for this reason that the scientific world of the day was so reluctant to recognize and acknowledge his achievement, deeming it incredible that such a man could succeed where they had failed. They misjudged their man. If birth and breeding play any part in success then the Devonian was as well qualified as any of those who danced attendance at the Court of St James or attended the meetings of the Royal Society.

The lineage of Thomas Newcomen can be traced straight back to Hugo le Newcomen, Lord of the Manor of Saltfleetby in Lincolnshire, who, in the twelfth century, accompanied Richard I on his crusade to the Holy Land. For nearly four hundred years an unbroken succession of Newcomens held their manor but then, like many another ancient family, political misfortune overtook them and they fell into obscurity. Bryan Newcomen was implicated in the Lincolnshire Rising of 1536 and for this his lands were confiscated by Henry VIII who awarded them to his favourite, Charles Brandon, Duke of Suffolk. The family thereafter dispersed to London, to Essex, to Ireland and to Devonshire. As though to set the seal upon the dissolution, in the next century Cromwell's soldiery were quartered at Saltfleet Manor and plundered the Newcomen memorial brasses in the church of Saltfleetby All Saints before they left the village to march to Lincoln.

Charles Newcomen, younger brother of the unfortunate Bryan, removed to London and it was his two sons, Elias and Robert, who founded respectively the Devonian and Irish branches of the family. Only the latter succeeded in retrieving the family's ancient station. Robert was knighted at Dublin Castle in 1605, and in 1625 he was created Sir Robert Newcomen, Baronet of Kenagh, Co Longford.

His descendants married daughters of the Earls of Mayo, Ross, Bessborough and Donegal while the fourth baronet married Ann Bullein, a collateral descendant of Henry VIII's unfortunate Queen. The nineteenth-century Viscount Newcomen, who died childless, was the last direct representative of this Irish branch, but there are many members of the distaff side both in Ireland and England today.

Elias Newcomen was born in 1547, matriculated at Clare College, Cambridge in 1565, obtained his MA at Magdalene College, Cambridge in 1572, and set up school in London. He translated from the Dutch an account of events in the Netherlands about 1575. Having joined the established church in 1588 he was appointed rector of Stoke Fleming, near Dartmouth on 19 March 1593/4 where he remained until his death on 13 July 1614. He married Protheza Shobridge of Shoreditch in 1579 when he was aged 32 and she was buried at Stoke Fleming 6 December 1621.

13 Cryptogram of E Newcomen R from the Church Register, 1538-1602 at Stoke Fleming.

The first Church Register of Stoke Fleming was beautifully transcribed by Elias during his ministry and commences with an ornate letter 'A' with tracery and a human face. There is a cryptogram or cipher signature which can be transcribed as E NEWCOMEN R. The last letter R, doubtless refers to Rector but the ER could perhaps be the royal ensignia since the transcript ended in the last year of Queen Elizabeth's reign and were required by one of her decrees.

Below this cryptogram are three lines headed 'Matt 19.5' correctly written in Greek, Latin and English, as one would expect of a classical scholar.

The church has two wall mounted brasses erected to the memory of Elias Newcomen; they

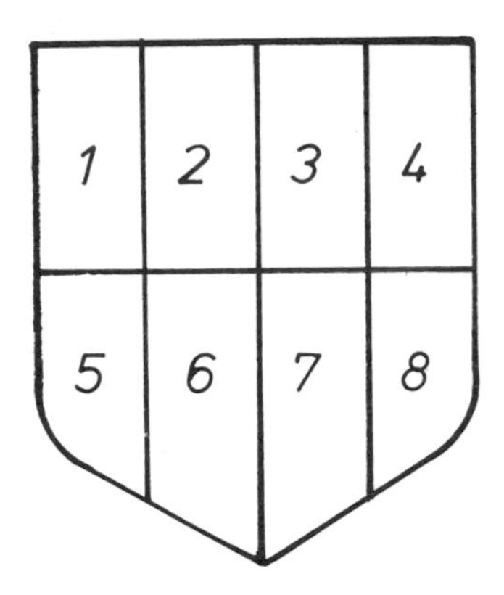

14 A copy of the brass in Stoke Fleming Church, Devon, to the memory of Elias Newcomen (1547-1614), Rector 1594-1614.

are thought to have been originally on the floor and to have included a figure.

The families denoted in the elaborate shield of arms have been recognised as (1) and (8) Newcomen, of Saltfleetby, Lincs; (2) King, of Gainsborough, Lincs; (3) Greenfield, of Barnebow, Lincs (his grandmother); (4) Stevenson, of Boston, Lincs; (5) Fereby, of Yorks; (6) Nightingale (the rose) of Brentford, Essex, (his mother's family); (7) Ellis (from whom she inherited her fortune). The inscription below the shield features the first of several puns on the family name:

Elias old lies here intomb'd in grave
but Newecomin to heavens habitation.
In knowledge old, in zeale, in life most grave,
too good for all who live in lamentation.
Whose sheep & seed with heavie plaint & mone
will say too late Elias old is gone.

The xiij of July 1614.

The rector's son, Thomas Newcomen, became a merchant venturer, a freeholder of Dartmouth, Receiver or Treasurer for the town during the Civil War and a staunch Parliamentarian. According to State Papers he was the owner of several ships and once claimed as a prize a vessel taken off the coast of Newfoundland.

A Chancery case in which he is described as 'Thomas Newcomyn of Dartmouth, Devon, Merchant' was heard in 1647. It concerns his share in a vessel *Desire* which was seized by a Parliamentary vessel while returning from Spain and taken to London as a prize. Later the ship was returned to Dartmouth but Newcomen sought compensation for loss of provisions and profits.

On his death in June 1653 he bequeathed to his wife Bathsheba and her children 'My ship *Nonsuch* with her guns, boats, etc and my interest in the ship once called *The Blessing*, now *The Nathaniel* of London'. Thomas also bequeathed the sum of '£62. 10s. 0d. adventured in the Irish acres of the Irish Rebel Lands' which suggests that there may have been a link between this Devonian merchant venturer and his Anglo-Irish first cousin, Sir Beverley Newcomen, who was 'Admiral for the Coasts of Ireland'. The change in the name of Thomas's ship is significant for there was then at Leyden an exiled nonconformist preacher named Nathaniel Newcomen. The inference is that it was at this time that the Devonshire Newcomens, led by the Parliamentarian Thomas, severed their allegiance to the established church.

It was Elias, the son of this Thomas who was the father of the inventor, Thomas Newcomen. He too, was a freeholder, a merchant of Dartmouth who died in 1702 and was buried on 4 August. His first wife was Sarah who was buried 24 March 1666/7 at Dartmouth. Shortly afterwards he married again. The ceremony was at the small church of Halwell on 6 January 1667/8 and his second bride was Alice Trenhale of Kingswear and she would have cared for young Thomas at Dartmouth.

Elias traded with the ship *Nonsuch,* which had been left to him by his father and was involved with other vessels. A Chancery case in 1677 tells of his interest in a new vessel, the *Mediteranean* which broke its mast during a maiden voyage from Plymouth to Newfoundland. This confirms that he was trading to distant parts and, in fact, was probably continuing his father's trade and business. Thomas, the inventor, reputedly his second son, would be already 13 years old at the time of this voyage. He was most probably born at the house on Dartmouth Quay which his family occupied for many years and until recently formed the southern part of the Criterion Restaurant and in 1976, Taylor's Café.

Newcomen, as we shall now simply call him, was christened at St Saviour's Church, Dartmouth, on 24 February 1663. The exact date of his birth is not known but, bearing in mind the fact that in those days the christening ceremony was never long postponed because of the high infant mortality rate, it is safe to say that he was born in late January or early February 1663.

Of his education and training there are no positive details. His father Elias was a member of the small group responsible for bringing the well-known nonconformist scholar, John Flavell of Bromsgrove, to Dartmouth in 1656 and it is more than likely that the boy was educated by Flavell. The importance of this link with Bromsgrove will appear later.

There is a very close link with the Baptist faith throughout the whole Newcomen story. Although there was no Baptist chapel at Dartmouth in Newcomen's time there were already meeting houses in Exeter in 1600, Plymouth in 1637 and in Kingsbridge in 1650.

Thomas Lidstone (1821-88), a Devonian architect and surveyor who became deeply interested in the story of Newcomen and wrote three pamphlets on the subject, states that Newcomen was subsequently apprenticed to an ironmonger in Exeter, but local research has failed to disclose the source of this information. Lidstone may have repeated an oral tradition which should not be dismissed as untrue because

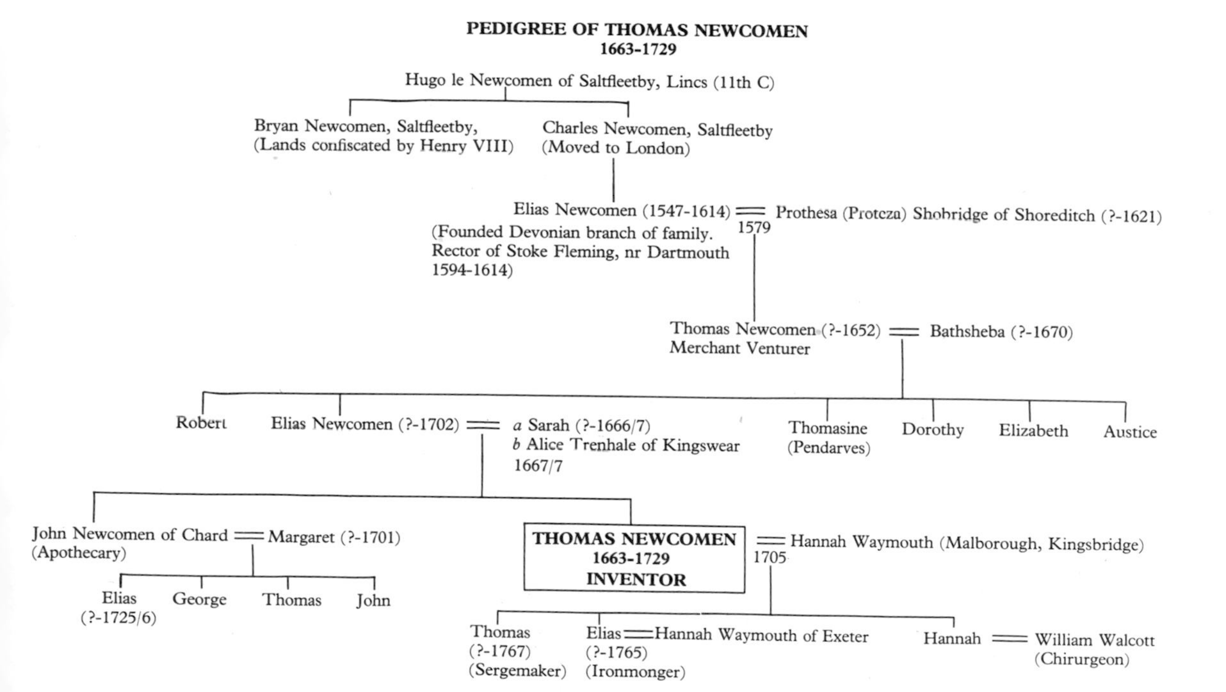

the Exeter city records have failed to confirm it.

If Newcomen did serve an apprenticeship in Exeter he would have returned to his native Dartmouth about 1685, to establish himself in business on his own account as an ironmonger. There is confirmatory evidence of this from 1688 in the Dartmouth Borough Receiver's accounts in which Newcomen is entered as supplying small items such as locks. Also it is shown that he was acting as Overseer to the Poor. Later, in the mayor's accounts he is recorded as supplying various small items such as latches and nails and also receiving payment for other items which are not specified. In 1704 he either supplied, or perhaps more likely repaired, the town clock for which he received £1 4s 1d, demonstrating his flair for mechanical contrivance.

These small items should not obscure the much larger general trade with which he was involved. The term 'ironmonger' at the period was used not merely for the retailer of small metal goods but for the inland merchant in the iron trade, making a proportion of the goods sold on his own premises. For example, Newcomen purchased between 1694 and 1700 varying quantities of iron produced by the famous Foley ironmasters on the River Stour in Worcestershire. In the year 1698/9 a total of nearly 25 tons was purchased and no doubt Newcomen would be dealing with other ironmongers and traders.

His partner at Dartmouth was John Calley whose name has sometimes been spelt Caley or Cawley. His marriage is apparently recorded in the Dartmouth registers on 8 October 1694 when John Cawley married Margery Towell. They moved to another house in a street leading to Newkay (now Newquay), Dartmouth in 1704, for which they paid 3s 4d quarterly in rent.

15 Signature of Thomas Newcomen from a Dartmouth house lease dated 23 Jan 1728/9, and signature of John Calley from the indenture of lease for his house at Dartmouth dated 2 July 1707.

Newcomen and Calley were both devout Baptists and the deeply religious tenor of a letter which Newcomen wrote to his wife only two years before his death shows that his nonconformist faith continued to be the mainspring of his conduct throughout his life. They are referred to by Farey and other authorities as Anabaptists and this is a common eighteenth-century term. There was no connection with the Continental Anabaptist movement of the fifteenth and sixteenth centuries.

In 1705 Newcomen, then aged 42, brought home a bride to Dartmouth. Said to be 19 years his junior, she was Hannah the daughter of Peter Waymouth, a farmer of Malborough, near Kingsbridge, doubtless from the Baptist community there. She bore him two sons, Thomas and Elias, and a daughter Hannah. Thomas became a sergemaker in St Mary Magdalene, Taunton, Somerset, while Elias later entered his father's business and remained at Dartmouth.

Newcomen's only daughter Hannah married a Mr Wolcot, the uncle of the John Wolcot, who, under the pen-name of 'Peter Pindar', became celebrated as a satirist. According to Lidstone, 'Peter Pindar' painted a portrait in oils of Newcomen which has since been lost, there being no known likeness of the great engineer. If such a portrait ever existed it could not have been painted from the life by John Wolcot as he was not born until 1738, nine years after Newcomen's death, and the story is probably apocryphal.

In 1707, Newcomen took the lease of a large house which extended from Higher Street to Lower Street in Dartmouth for which he paid rent of £1 6s 8d per year. Part of this house he assigned to his Baptist friends as a place of worship and for some years acted as their leader and teacher. Later, when business took him away from Dartmouth, it is said he continued to preach to fellow Baptists elsewhere. Despite its size, the low-ceilinged, timber-framed construction of Newcomen's house makes it unlikely that it was the scene of the protracted experiments which led to the birth of the steam engine.

He had a garden on the south side of Ford Lane and there were also some cellars or ground rooms in Dark Lane and a small tenement which adjoined the Guildhall. These may have been the

16 A drawing of Newcomen's house at Dartmouth (centre left) before demolition in 1864. Sloping downpipes take water from the roof.

site of the workshops where he and Calley were to experiment.

In 1841, A. H. Holdsworth, the governor of Dartmouth Castle, published a pamphlet entitled *Dartmouth, the Advantages of its Harbour as a Station for Foreign Mail Packets* in which he wrote: 'As I knew the house in which Newcomen lived in Dartmouth I purchased the carved panels of his sitting-room, and have placed them in a room of my own at Brookhill as the best existing record of a man to whom England is so greatly indebted'. This suggests that the old house had by then fallen on evil days. Holdsworth installed the linenfold panelling in an octagonal summer-house at Brookhill, Kingswear, where it remained until 1955 when it was removed to the Borough Museum at the Butterwalk, Dartmouth.

In 1864, Newcomen's house was demolished in order to widen the Higher and Lower Streets and to construct a ramp between them. All that now remains is a commemorative plaque built into the ramp retaining wall, but fortunately the appearance of the house was recorded. A sketch of the Lower Street façade was reproduced in a pamphlet by Thomas Lidstone entitled *Some Account of the Residence of the Inventor of the Steam Engine,* published in 1869. This façade also features in an early photograph which R.P. Leitch clearly used when making the drawing which appears in Samuel Smiles's *Lives of Boulton and Watt.* Smiles refers in a footnote to timbers supporting the gable ends and states that the house was demolished because it had become unsafe. Evidently Smiles never saw the house itself or the photograph for, as the latter makes clear, what he took to be timber props were in fact down-spouts. Timber-framed houses of similar type survive in Dartmouth to this day so it is possible that street widening and not decay was the main reason for demolition.

When the old house was pulled down, the indefatigable Lidstone purchased the elaborately carved timberwork of the Lower Street façade and incorporated it, together with materials from other demolished houses, in 'Newcomin Cottage', a house which he built for himself at Ridge Hill, Dartmouth, in 1868. Over the fireplace in the large first floor living-room of this house (now divided into two bedrooms) Lidstone set the original oak clavel or lintel from the hearth where, according to the local folk tale, recorded by Holdsworth, Newcomen 'discovered' the power of steam by observing a boiling kettle.

Newcomen's assistant, John Calley was a plumber and glazier while Newcomen himself is described by some early writers as a blacksmith, an appelation which tends to denigrate and gives an indication of only some part of his work. We may be sure that when Newcomen visited the mines of Cornwall and Devon, as we are told he frequently did, it was not just to sell but as a practical knowledgeable engineer craftsman. To the end of his days he was proud to describe himself as 'Ironmonger'.

An insight into Newcomen's business activities is given in a legal suit which, although it was commenced late in his life, refers to an earlier period. In evidence Newcomen stated that from 1702 (which was the date of his father's death) 'having occasion for money to carry on his affairs did several times borrow and take up at interest several sums of money'. Clearly his own finances were insufficient to meet his business demands even from this early date.

To satisfy the multifarious needs of their customers Newcomen and Calley had to be adept at every variety of metal work in iron, brass, copper, tin and lead. Add to this that Newcomen was a member of a respected family with a long background of world trading and it is possible to appreciate how the two men possessed precisely the right combination of commercial ability and practical, versatile craftsmanship, which was required to bring the steam engine to a successful birth. Newcomen's predecessors had plenty of intelligence, but in the second and no less vital qualification they had all been woefully deficient.

Because of the lack of solid facts about the birth of Newcomen's great invention, his biographer can do no more than choose the most likely and credible path through the maze of legend and conjecture which earlier writers have created. When Newcomen's mind first turned to the steam engine there can be no knowing, but it is safe to say that he was prompted by his frequent visits to the tin mines which gave him an intimate knowledge of their pumping problems. Surely, too, he would be aware of the drainage problems which existed in the coal mining areas of the Midlands and North East.

The most interesting question is to what extent, if at all, Newcomen, when he first embarked upon his experiments profited by the work of his predecessors and of his contemporaries, Papin and Savery. Because it seemed to the scientific writers of the eighteenth century inconceivable that a provincial ironmonger could succeed unaided where the scientific world of London had failed, they postulated links between Newcomen and that world.

An assumption was initiated by Dr John Robison, the friend of James Watt, in his article on 'Steam Engines' in the third edition of the *Encyclopaedia Britannica* (1797). Robison claimed to have found documents in the archives of the Royal Society which proved that Dr Robert Hooke was in communiction with Newcomen, that he informed the latter about Papin's work and suggested the direction which his experiments should take. This story has since been frequently repeated by later writers but it has now been discredited by the two greatest authorities on the subject, Rhys Jenkins and H. W. Dickinson. Dr Dickinson could find no trace of such documents or references among the minutes and records of the Royal Society which were bound up in the mid-eighteenth century and are therefore in exactly the same state as they were when Robison, who was not a member of the society and therefore enjoyed no special privileges, consulted them and moreover, Robison's article abounds in errors of a very careless and obvious kind. The present conclusion is that, if Newcomen had any knowledge of Papin's work, it was through the account of his experiments which appeared in *Philosophical Transactions,* but although this is perhaps unlikely, it is important not to underestimate the degree of communication that took place at this time. Letters could be sent by regular coach services and deliveries not much worse than present day services were regularly achieved. Dartmouth had for centuries been a most important port and was well served with communications by land and sea. Newcomen's trading with the Midlands has already been shown and no doubt he was in contact with other areas.

Stonier Parrott who was to play a considerable part in the uptake of the invention was in 1725 concerned to avoid the payment of licence fees and proposed that the patent should be challenged. In his memorandum, first quoted by E. Hughes in *Archaeologia Aeliana,* he claims that Newcomen's engine 'was invented by Monsieur Pappein and others, as I found sufficiently described in their writings from which Mr Newcomen began to make improvements'. Although Parrott was being deliberately supercilious in his memorandum, he had known both Savery and Newcomen.

It has been said that Newcomen saw a Savery engine at work in the Cornish mines, that Savery and Newcomen were well known to each other and that Savery himself described Newcomen's engine as an improvement upon his. Farey says as much in his *Treatise on the Steam Engine,* being among those who claim that Savery acquired his title of 'Captain' through long association with the Cornish miners. Savery may have installed one of

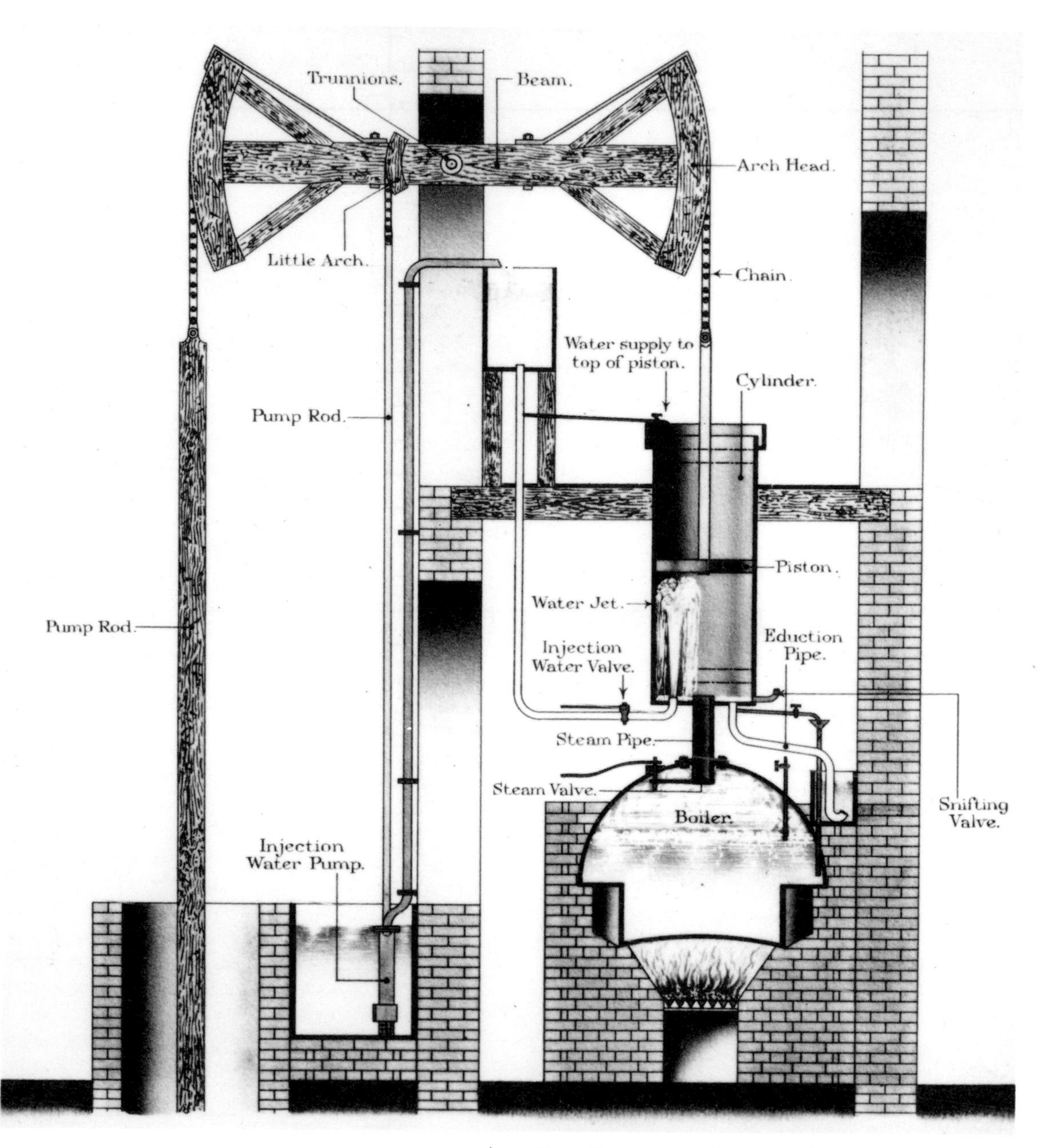

17 A diagram showing the principal features of the Newcomen engine in section. Steam is generated at atmospheric pressure in the boiler and fills the cylinder during the upward stroke of the piston. The steam valve is then closed and the steam is condensed by a jet of cold water causing a vacuum under the piston. The atmospheric pressure acting on the top of the piston forces it down, hence the description 'atmospheric' engine, and this constitutes the working stroke. The piston is raised again by the overbalancing weight of the pump rods.

his steam pumps in Cornwall which Newcomen could have seen, but there is no positive evidence of this.

It is now known that more positive opportunities existed for Newcomen and Savery to meet. Savery's duties as treasurer of the Sick & Wounded Commission from 1705 required him to visit the various ports and stations where he had to pay the commissions of the agents. From Savery's own accounts we find that on 9 January 1706/7 he was at Dartmouth visiting the mayor, Caleb Rockett, who was the agent. These visits were on a regular basis and it was not until March 1712/13 that the agents were required to visit London for payment rather than the treasurer, Savery, should visit them. Newcomen's name has already been noted as appearing in the mayor's accounts and so here is positive evidence that both Savery and Newcomen were known to Rockett at Dartmouth where they could well have met.

When Savery joined the Royal Society in 1705 he took an active part in its affairs. It was on 11 February 1707 that Papin presented to the society his paper on an 80-ton steamship and Savery's reply 'on a pretended invention of Dr Papin' was on 18 May 1709.

Savery's visits to Dartmouth were during the very time of Newcomen's intense experimental activity and Savery's interest in the subject. According to the accounts of Triewald and Switzer, Newcomen may have already completed the basic design of his engine by this time and the partnership in the patent may have been discussed during these visits to Dartmouth.

It will be remembered that what Papin did with his classic experimental apparatus was to generate steam in the bottom of a cylinder below a piston and then condense it by allowing the cylinder to cool, thus creating a vacuum. What Newcomen did was to combine Guericke's cylinder and piston with Savery's separate boiler. He placed his boiler beneath the vertical cylinder and connected its piston to one end of an overhead beam or 'great lever'. Below the piston steam was condensed by internal injection creating a vacuum by which the piston descended to actuate the beam and lift the pump rods connected to its opposite end. If Newcomen did see a Savery pump at work in Cornwall, the boiler would be the only feature to interest him; the rest would simply be a lesson in what to avoid. Newcomen the practical craftsman would surely see that Savery's use of high-pressure steam was beyond the technical competence of the time and he therefore very wisely decided to confine his efforts to harnessing the power of atmospheric pressure alone. When the result of these efforts finally emerged it was no mere improvement on Savery's machine but something entirely new, an immense advance upon anything which had gone before. It represented for the first time the classic combination of major components which, with numerous refinements, would persist for nearly two hundred years.

In the whole history of technology it would be difficult to find a greater single advance than this and certainly not one more pregnant with significance for all humanity. It is no wonder that there were many who doubted whether a provincial ironmonger could possibly have been sole author of so momentous an invention. It is certainly tempting to see it as a development of Papin's experiments, yet history shows that there is nearly always parallelism in invention because, when a technical problem becomes sufficiently pressing, several minds may be simultaneously and independently bent upon finding a solution. Newcomen hit upon the right solution precisely because he was a practical craftsman and so evolved a machine whose construction was strictly within the competence of the craftsmen of the day.

In support of the belief that Newcomen owed little or nothing to Papin or Savery the young Swedish engineer Marten Triewald is the most convincing witness. Triewald came to England in 1716 when he was twenty-five to study English industrial methods and he did not return to his home country until 1726. A highly intelligent man, he harboured none of the prejudices which led the English scientists to denigrate Newcomen and his engine. He met Newcomen and his assistant Calley and was immensely impressed by their engine. So much so that he subsequently assisted in the erection of a Newcomen engine at Byker Colliery near Newcastle, while after he returned to Sweden he was responsible for building an engine at the Dannemora Mines. In 1734 he published in Stockholm a book whose translated title is, *A short description of the Fire and*

Air Engine at the Dannemora Mines, in which he covers the history of the Newcomen engine down to that date. This did not become available in English translation until 1928 but the extracts which follow are from a revised translation by Torsten Berg in 1975. It is reasonable to attach more weight to Triewald's testimony than to that of authors such as Robison who wrote long after the date and who, moreover, had not read Triewald.

After a preliminary paragraph about Thomas Savery and his engine, Triewald writes as follows:

> Now it happened that a man from Dartmouth, named Thomas Newcomen, who had no knowledge whatsoever of the ideas of Captain Savery, had at the same time also made up his mind, in conjunction with his assistant, a plumber by the name of Calley, to invent a fire-engine for pumping water from the mines. He was induced to undertake this by considering the heavy costs of drawing water by means of horses, which he found in use at the English tin mines. Mr Newcomen often visited these mines in the capacity of a dealer in iron tools which he used to supply to many of the tin mines.

This account sounds entirely logical and convincing, suggesting that it was written by a man who was sure of his facts. Furthermore, it is corroborated by another early writer, Stephen Switzer, who, in his *Hydrostaticks and Hydraulicks* (1729) had this to say:

> I am well informed that Mr Newcomen was as early in his Invention as Mr Savery was in his, only the latter being nearer the Court, had obtain'd his Patent before the other knew of it, on which account Mr Newcomen was glad to come in a Partner to it.

In the face of all the evidence, the author is inclined to the view that Newcomen was aware of Savery's ideas and his engine and that they may well have met by 1705.

Since Newcomen's engine was totally different from Savery's pump, the Patent Law at the time would have allowed him to take out his own Patent, but he could not have operated his engines without infringing the Savery patent. Since we do not know exactly when Newcomen joined Savery or indeed on what terms, it is difficult to be certain about the considerations which led to this co-operation. The assumption has commonly been made that Savery, having a prior patent widely and vaguely worded, was able, possibly by the threat of action at law, to make a deal with Newcomen and force him to sell his engine only at the cost of some payment to Savery or his representatives. If this were so, given the utterly different commerical possibilities of the two engines, Newcomen might indeed have laboured under a grave injustice. Certainly we have evidence that Savery fought hard for the claims of his inventions and is unlikely to have passed by any supposed infringement without opposition or public protest.

We can appreciate this, not only from the *Miner's Friend* but from his *Navigation Improved or the Art of Rowing Ships of All Rates, etc* (1698) in which he recounts his arguments with the Navy over the efficacy of an invention patented in 1696. A different interpretation of the relationship of the two steam engine inventors is possible, and arguably more likely. Despite the claims of the *Miner's Friend* it must soon have become apparent that Savery's engine had no practical value in mining and that the purpose of the extension of the patent by Act of Parliament in order to enable him to further improve the device was not likely to be fulfilled, for in 10 & 11 Wm III Cap 61 (1699) it was said that Savery 'may and probably will require many years' time and much greater expense than originally hath been to bring the same [engine] to full perfection' which is to be contrasted with his excessive claims in the *Miner's Friend*, written towards the end of 1701, that the engine was 'fully compleated and put into practice in your dominions, with . . . repeated success and applause'.

Newcomen, with his practicable engine, would have had to get his own patent to protect it, possibly in the face of costly opposition from Savery. If he succeeded, he would only have the normal fourteen year period of monopoly unless he too was prepared to go to the expense of obtaining an Act of Parliament to extend his patent. The very ill-success of Savery's engine might have militated against the exceptional passing of an extending Act for a second invention of apparently a very similar kind. But by an arrangement with Savery, Newcomen and his associates could gain real advantages. They avoided the expense of obtaining a patent themselves. They obtained, even if the agreement only became operative towards the end of Savery's life, an increased number of years of monopoly above the fourteen given under a normal patent, without the expense, or the risk of rejection, of a Bill of their own. If in fact an understanding between Savery and Newcomen was arrived at much earlier, just possibly as early as 1705, then Newcomen could have enjoyed a long period of protection covering much of the inevitable developmental period as well as that period of commercial viability under a patent which produced the astonishing sale figure of at least 100 engines.

Newcomen then, had very real advantages to obtain from the shelter of Savery's extended patent, and he and his associates may well have been willing to pay a handsome consideration for this facility, while Savery may have been prepared to accept the arrangement because he could share in the licence fees paid by customers for an engine which had much better prospects of sales than his own. Newcomen could have received a patent for his engine during the period in which Savery's was in force and Triewald in fact patented a modified Newcomen engine in 1722 during the period of Savery's patent. This may be a further indication that Newcomen's association with Savery was voluntary and calculated, not unwilling and disadvantageous. Finally, contemporaries can be shown to have believed that the Newcomen associates were getting an improper advantage from the protection of the extended Savery patent. Stonier Parrott (although not always the most trustworthy of men) in proposing a petition against this arrangement in 1725 was on sound ground in stating that it was wrong that two very dissimilar machines should enjoy the protection of the same patent when they were as different 'as a Distil is from a Windmill'. He even claimed that Savery had never been able to understand Newcomen's engine — 'to his dying day I could never make him understand how that Engine was wrought or to have any opinion of it'. At present we must admit that we lack sound documentary evidence for the exact relationship between Savery and Newcomen, but it would be unjustified to suppose that the association was simply one between the exploiter and the exploited, and we should be on the look out for evidence which might illuminate the origins and nature of their co-operation.

The first positively recorded Newcomen engine is dated 1712 which means, if we accept the statements of Triewald and Switzer, that Newcomen devoted many years to the development of his engine before he achieved practical success. What went on in the little Dartmouth workshop during these long years of gestation we do not know. Of the endless, patient experiments, the trials and the repeated disappointments there is no record but it is possible to advance some plausible conjectures. In the first place, it is obvious that this experimental period must have been greatly prolonged by the fact that work had to be confined to limited spare time, since Newcomen and Calley could not afford to neglect their day-to-day business. It is also safe to say that Newcomen's first attempts to give his ideas a practical shape would consist of small models. To build anything larger would have been beyond the scope of the means available in so modest a workshop where small brass castings were probably the most ambitious product. Although it is extremely doubtful whether Newcomen would have realised it, in experimenting with such small prototypes he would have been greatly hampered by what the modern model-maker calls 'scale effect'. An accurate scale model will not work as efficiently as its large prototype because, relatively, frictional and thermal losses increase as the size diminishes. As every model engineer knows, scale effect is very marked even though the chosen prototype may be well designed and highly efficient. Where the efficiency of the prototype is low the scale effect may be so crippling that the model will not work at all. The young James Watt discovered this when, as an instrument maker, he was asked to repair a demonstration model of a Newcomen engine for the University of Glasgow. Try as he would he could not get the model to work for more than a few strokes before the boiler ran short of steam. Watt might never have embarked upon the research which led directly to his famous invention of the separate condenser had not the model exaggerated for him the inefficiency of the Newcomen design. When we remember that Watt worked on the Newcomen model in 1763-4, when the Newcomen engine was firmly established in commercial use and had been much improved by Newcomen's successors, it is easy to imagine the long, disheartening struggle which Newcomen and Calley must have waged with their primitive apparatus in the Dartmouth workshop.

It seems likely that for his early experiments Newcomen used a cast brass cylinder of a diameter not exceeding 7in as this was the size commonly used for pumps at that time. Even so, no machine existed which could bore it truly cylindrical so that it needed to be laboriously fettled and lapped by hand. In these circumstances it was extremely difficult to obtain a well-fitting piston. Again following pump practice, a leather-faced piston was used but in addition Newcomen introduced a water seal, allowing a small quantity of water to lie upon the upper side of the piston. This seal could be replenished as needful since the upper end of the cylinder was open to the atmosphere.

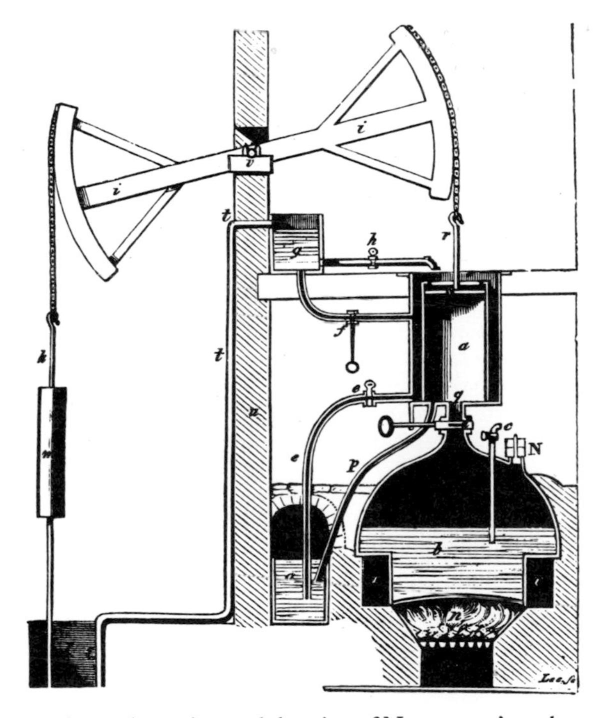

18 A purely conjectural drawing of Newcomen's early design showing a jacketed cylinder and hand-operated valves. From Stuart's *Descriptive History of the Steam Engine,* 1828. The drawing is anachronistic and shows a much later design of boiler.

It will be remembered that in Papin's experimental apparatus, when the cylinder had been filled with steam it was allowed to cool down naturally, thus condensing the steam within and so creating the vacuum which enabled atmospheric pressure to drive the piston down. Whether or not he knew of Papin's experiment, it must very soon have become obvious to Newcomen that before an engine on this principle could be considered a practical proposition some more rapid method of condensing the steam was essential. Otherwise the working cycle would be quite intolerably slow. According to Desaguliers and other writers, Newcomen's first step was to try cooling the cylinder walls externally by means of a flow of cold water on the same principle that Savery adopted to cool the receivers of his pump. But whereas Savery could allow water to flow freely from the rising main over his closed receiver this crude method was not applicable to Newcomen's open-ended cylinder. Consequently it seems clear that Newcomen enclosed the sides of his brass cylinder in a lead jacket and circulated cold water through the space between them. Stuart, in his *Descriptive History of the Steam Engine,* illustrates what purports to be Newcomen's first engine. This shows the cylinder jacket supplied with water from the same cistern which replenishes the water seal on the top of the piston. A drain pipe from the jacket leads to the hot well into which the cylinder condensate discharges. According to Stuart the jacket water was drained into the hot well and replaced with cold water from the overhead cistern after each power stroke. Hand-operated valves are shown. This was an improvement, but the action would still be far too slow. The next development was to introduce a jet of cold water into the steam-filled cylinder. This produced the desired result — far more rapid and effective condensation — and it was this truly dramatic advance which made the engine a practicable proposition.

Whether this decisive step forward was the fruit of patient experiment or whether a fortunate accident pointed the way has been the subject of much debate. Desaguliers held that the discovery was accidental and his account of what took place reads as follows:

> One thing is very remarkable. As they at first were working they were surprised to see the engine go several strokes and very quick together, when after a search they found a hole in the piston, which let the cold water [from the water seal] in to condense the steam in the inside of the cylinder, whereas before they had always done it on the outside.

This account led H. W. Dickinson to dismiss the theory of fortunate chance altogether because it is obviously fallacious. It is inconceivable that the engine could have made those 'several strokes very quick together' with a hole in the piston which would have admitted water or air continuously and so prevented the creation of a vacuum. Again, the quantity of water maintained above the piston was small and it would probably be hot. However, if we turn once again to the evidence of Marten Triewald it is difficult to resist the conclusion that the great discovery *was* the result of a fortunate accident but that the exact circumstances were wrongly reported by Desaguliers and other writers after him. This is Triewald's account which he may well have had from Newcomen's own lips:

> The work on the prototype of the fire-engine was carried on for ten years altogether and may never have achieved the desired result, if Almighty God had not providentially allowed something very special to occur. During the last test that they intended making with the prototype it happened that a greater effect than expected was caused by the following event. The cold water, that had been allowed to flow into the leaden jacket, surrounding the cylinder, penetrated the cylinder wall through a casting fault mended with tin-solder, which had been melted by the steam. Forcing itself into the cylinder the cold water immediately condensed the steam, causing such a high vacuum that the weight hooked on to the one end of the little beam of the prototype, which was supposed to represent the weight of water in the pumps, was so insufficient and the air pressed with such a tremendous force on the piston that its chain broke and the piston itself knocked the bottom of the cylinder out and smashed the lid of the small boiler. The hot water flowing everywhere convinced also their senses, that they had discovered an incomparably powerful force, that hitherto had been entirely unknown in nature — at least the way in which it originated was unknown.
>
> Though some might think that this was an accident, I for my part, find it impossible to believe otherwise than what happened was due to a very special ordering of Providence, particularly when I, who knew the original inventors, consider that the Almighty then bestowed upon mankind one of the most wonderful inventions ever brought into the light of day, and this through persons who had never acquired higher learning or academic degrees.

The sequence of events here described in such circumstantial detail seems to this writer perfectly convincing. A small blow-hole in the brass cylinder has been stopped with solder. When this stopping fails there is sufficient volume and head of cold water in the surrounding jacket to produce a strong jet of water through the hole and into the steam-filled cylinder. Steam at no more than atmospheric pressure would not have been hot enough to melt the solder as Triewald states, but that the flux had not 'taken' in the blow-hole is more than likely. The engine could no more continue running with a hole in the cylinder wall than it could with a hole in the piston but, unlike Desaguliers, Triewald tells us clearly that it did not do so. It made a single stroke so violent that it broke the chain connection on the pump end of the beam and damaged both the cylinder bottom and the top plates of the boiler beneath it.

Desaguliers belonged to that scientific school which refused to credit that a practical man without scientific training could achieve anything except by chance and good luck. Therefore, according to him, the development of the steam engine was simply one long, phenomenal run of luck. This is how he puts it:

> If the Reader is not acquainted with the History of the several Improvements of the Fire-Engine since Mr. Newcomen and Mr. Cawley first made it go with a Piston, he will imagine that it must be owing to great Sagacity, and a thorough Knowledge of Philosophy, that such proper Remedies for the Inconveniences and difficult Cases mention'd were thought of; But here has been no such thing; almost every Improvement has been owing to Chance.

Thus the pundits attempted to belittle Newcomen's achievement, but to accept Triewald's account of the chance discovery of internal injection is not to play Desaguliers's game. Such fortunate accidents do sometimes happen to aid the inventor, but by the law of averages they are rare and therefore they only reward the man of infinite tenacity and patience. There is never a short cut to success. Newcomen earned his reward, for on Triewald's admission he had laboured fruitlessly for ten years before the problem of rapid condensation was solved for him. There must have been great rejoicing in the little Dartmouth workshop on that day, but if the way ahead was now clearly signposted, there were still some problems left for Newcomen to solve and they were not solved by chance: how to rid the cylinder of hot condensate and of the air which entered it with the steam, and would destroy the vacuum if allowed to accumulate; how to make the steam admission and water injection valves self-acting. By the time the first recorded full-sized Newcomen engine was built in 1712 these problems had been triumphantly surmounted. A new power was launched upon the world.

CHAPTER 3

The First Newcomen Engines c1710-15

As we shall see presently, the design of Newcomen's engine of 1712 was so mature that it is incredible to assume, merely because of the lack of evidence, that there was no intermediate stage of development between it and the small workshop model described by Marten Triewald. Stories are told by Stuart and others of early Newcomen engines with hand-operated steam and injection valves and if there is any truth in these at all it is almost certain that they apply to an engine or engines built before 1712.

For geographical reasons we should expect to find Newcomen making the first full-scale trial of his engine in the mining districts of Devon or Cornwall where he was well known. Tin mining in Devon, in the neighbourhoods of Ashburton, Tavistock and Widecombe on the edge of Dartmoor, was pursued on a very small scale compared with Cornwall, but from Newcomen's point of view these mines had the advantage of close proximity to Dartmouth. Although tin output in Devon reached its peak at the end of the fifteenth century and the years immediately following, it is perhaps significant that there was a brief revival during the years 1705 to 1711. The highest output of tin was reached in 1706, but in 1712 production slumped to precisely half the previous year's total and never again recovered. This sudden slump could have been due to the drowning of a mine owing to the failure of an experimental pumping engine. Even if this were not the case it would provide Newcomen with an additional reason for looking further afield for a market for his invention, but he would have been very much aware of the demand in the mining areas. There is certainly no record or legend of any early engine in Devon, but evidence could still come to light.

Joseph Carne in his paper on the history of copper mining in Cornwall quoted by R. L. Galloway in *The Steam Engine and Its Inventors* states that the first engine in Cornwall was erected at Wheal Vor tin mine in Breage where it worked from 1710 to 1714. This statement is supported by Hamilton Jenkin, in *The Cornish Miner*, who adds that the pumps at Wheal Vor were arranged in six lifts of ten fathoms each, a separate rod or 'spear' for each pump being hung from the engine beam. The discovery of a six-holed spear guide in the mine in 1815 is said to confirm this, but it is inconceivable that an engine installed at so early a date could have coped with such a load of rods and pumps.

Jenkin also quotes a 'vague tradition' of an even earlier engine first installed at Balcoath Mine, in the parish of Wendron near Helston, where turf fuel was used to raise steam, and subsequently moved, first to Tregonebris and finally to Trevenen Mine in the same neighbourhood.

The association between Newcomen and Savery not unnaturally led to considerable confusion between the works of the two engineers and it has been suggested that Wheal Vor may have been the site of the pump which Savery is alleged to have erected in Cornwall. Professor William Pole believed that Wheal Vor was the scene of one of Newcomen's first attempts and there is evidence to suggest that he was right. In the first place, if Savery did erect one of his pumps in Cornwall it would have been earlier than 1710 since we are told that he abandoned his attempts to introduce his engine in mines in 1705. This date is significant because it is the same year in which Newcomen is said to have entered into partnership with him. Again, the Proprietors who acquired Savery's patent rights after his death claimed in 1716 that they had an engine at work in Cornwall whereas it was not until 1719-20 that the first Cornish Newcomen engine of which there is positive record was erected at Wheal Fortune mine at Ludgvan, near Penzance. Finally there is the testimony of Davies Gilbert (1767-1839), the friend of Richard Trevithick, and an eminent engineer, who in *The Parochial History of Cornwall* (1838) which he edited, wrote as follows:

> The steam engine, which consists essentially in a piston alternately sliding through a cylindrical vessel, invented by Mr Newcomen of Dartmouth had been used in one mine called the Great Work in Breage,

> when Mr Lemon came forward . . . commenced working a mine on a farm called Trowel, in the parish of Ludgvan, the property of Lord Godolphin, and named Wheal Fortune, where the second steam engine was used.

'Great Work' equals the Cornish 'Huel Vor' or 'Wheal Vor'.

Added together, this evidence makes a good case for a Newcomen engine at Wheal Vor in 1710. This engine would have had water injection to the cylinder, because the engine which Triewald described was obviously a small experimental model of earlier date with a weight simulating the pump load, but that it had self-acting valve gear is questionable. If Carne's statement is correct that the engine only worked for four years then it can be assumed that either it was not satisfactory or the mine failed and the ensuing events assume a logical sequence. Disillusioned by the failure at Wheal Vor and possibly other even earlier experiments, the Cornish mines captains, never an easy body of men to deal with as James Watt discovered years later, lost interest in Newcomen and his new fire-engine. One must remember that in Cornwall the engine would be most expensive to fuel and this would explain why Newcomen sought and achieved success, not on his 'home-ground' but in the Midlands where the high coal consumption of his engine was not such a drawback. It also explains why ten years elapsed before a second engine was erected (at Wheal Fortune) for by that time sufficient engines were at work elsewhere to convince the most sceptical Cornishmen that Newcomen's invention was a practical proposition.

The invasion of the Midlands by Newcomen and Calley is described in some detail by Desaguliers. It has always been understood that he relied on information from Henry Beighton, another man of science and a sceptic of the ability of Newcomen and Calley. Since Desaguliers published his book some years after the introduction of the engine, perhaps misunderstanding and errors may therefore have crept into his account. Recently however, letters written in French by John O'Kelly in 1721 to one of his business associates then in Vienna confirm that much of what is quoted by Desaguliers of the introduction of the engine must have come from a written source of about 1718-20 and that the author was almost certainly Henry Beighton. John O'Kelly (1672-1753) a member of a landed Irish family was born in Galway and was a Captain of Service in England in 1711. His first wife was Catherine Leland and his second, whom he married in Brussels 30 December 1723, was the Noble Marie Albine Angelique Van der Moere, daughter of the Count Jean de Rochefort.

O'Kelly must have first gained experience of the engine in England and then became engaged to build an engine in Belgium in 1720. In a letter 17 March 1721 attempting to confirm his knowledge of the engine he writes, 'Please ask Mr Potter if Mr Beighton understands, and if he confirms this, please rest assured that I also understand it'. Isaac Potter was building an engine near Vienna at this time and it becomes clear that both he and O'Kelly had been involved together with Beighton in England.

O'Kelly's letter refers to the Marquis of Worcester's engine and then describes the Savery engine — 'but it was not a machine such as that used by Mr Potter and myself today, since the machine built by Mr Savery cannot draw water lower than atmospheric pressure will raise it'.

After referring to Newcomen and Calley's invention he states that

> towards the end of 1711, they put forward proposals to draw the water from Griff in Warwickshire but these were not accepted, but by the intermediacy of Mr Potter in Bromsgrove, Worcester (as I believe father of the one at present in Vienna) they undertook to draw water for Mr Back of Wolverhampton, where after several trials, they accidentally succeeded, since neither is sufficient of a philosopher or mathematician to explain what they did, nor to calculate their proportions.

Desaguliers, clearly using the same document as a basis, repeats and further elaborates the somewhat disparaging story.

> . . . In the latter End of the Year 1711 [they] made Proposals to draw the Water at Griff in Warwickshire; but their Invention, meeting not with Reception, in March following, thro' the Acquaintance of Mr Potter of Bromsgrove in Worcestershire, they bargain'd to draw water for Mr Back of Wolverhampton, where, after a great many laborious Attempts, they did make the engine work; but not being either Philosophers to understand the reasons,

or Mathematicians enough to calculate the Powers, and to proportion the Parts, very luckily by Accident found what they sought for. They were at a loss about the Pumps, but being so near Birmingham, and having the assistance of so many admirable and ingenious Workmen, they soon came to the Method of making the Pump-Valves, Clacks and Buckets; whereas they had but an imperfect notion of them before.

This is another example of the scientists' refusal to credit Newcomen's success to anything other than chance and good luck. The statement that Newcomen and Calley 'were at a loss about the Pumps' is another piece of nonsense. No Midlander could teach men familiar with conditions in the Cornish mines anything about mine pumping, but in executing their ideas they undoubtedly would benefit very greatly from the metal-working craftsmen in that area.

The Beighton/Desaguliers account of the inception of the first successful Newcomen engine may be compared with that of Marten Triewald. Having described how Newcomen and Calley were compelled to join forces with Savery because the latter held a master patent, Triewald continues:

. . . Later on Mr Newcomen built the first fire-engine in England in the year 1712, which erection took place at Dudley Castle, in Staffordshire.
The cylinder of this engine measured 21 inches in diameter, and was 7 feet 10 inches high. The boiler was 5 feet 6 inches in diameter and 6 feet 1 inch high. The water in the boiler stood 4 feet 4 inches high and the volume of the water was 673 gallons. The machine made 12 strokes a minute and delivered 10 English gallons a stroke. The mine was 51 yards or 25½ fathoms deep.
The fame of this excellent pumping-engine soon spread across England and many people came to see it, both from England and from foreign nations. All of them wanted to make use of the invention at their own mines and exerted themselves to acquire the knowledge needed to make and erect such a wonderful engine, but the inventors, Newcomen and Calley, were exceedingly jealous and very anxious to preserve exclusively for themselves and their children, the knowledge of making and operating their invention, which had cost them such unprecedented toil. Even the Spanish Ambassador to the Court of St James's who came from London with a large suite of foreigners, to see the new engine, was not even allowed to enter the engine house however large a reward he offered and had to return very dissatisfied without having seen anything but the wonderful results, that this small engine (compared to that at Dannemora) was able to produce.

Thus Triewald gives a specific account of the engine and there is no doubt that it would arouse very widespread interest amongst the mine owners and operators throughout the country. The rapid take up in widely dispersed areas of the country is proof enough of the spread of interest, although in common with many other important inventions not a great deal was published at the time of the invention. Mine owners and operators were not necessarily interested in the publication of books but the general public's interest was such that Thomas Barney and Sutton Nicholls produced drawings and descriptions for sale in 1719 and 1725.

That Newcomen and Calley were jealous of their invention may well be true. The piracy of inventions was difficult to prevent in those days and at the outset their mode of proceeding — presumably with the assent of Savery — appears to have been, not to grant licences, but to hire out their services as drainers of mines by the aid of their engine.

The references by Triewald and Desaguliers, both published in 1734, are the earliest published histories of the subject other than J. F. Weidler (1728) and S. Switzer (1729) neither of whom comment on the exact location of the first engine. Two positive references however occur in 1725, the first an entry in the Bilston Parish Register, which reads as follows:

1725 Apr. 8. Thomas ye Son of John Hilditch Mangr of Ye Fire Engine att Tipton (he gives yt his Settlement is Bartomly – p'ish, Cheshire).

The second is an entry in the Diary of John Kelsall who visited the Tipton-Dudley district on 21 September 1725 and wrote 'There is near Dudley Castle an Extraordinary Fire Engine that throws up (as I remember) about 60 Hogsheads of

19 Newcomen's engine of 1712, as engraved by Thomas Barney in 1719. Built near Dudley Castle on part of the Coneygree Park or Tipton Field just within the boundary of Tipton, Staffordshire. The engine was moved to the Level Coal work, Dudley about 1740, later to Willingsworth Coal Works.

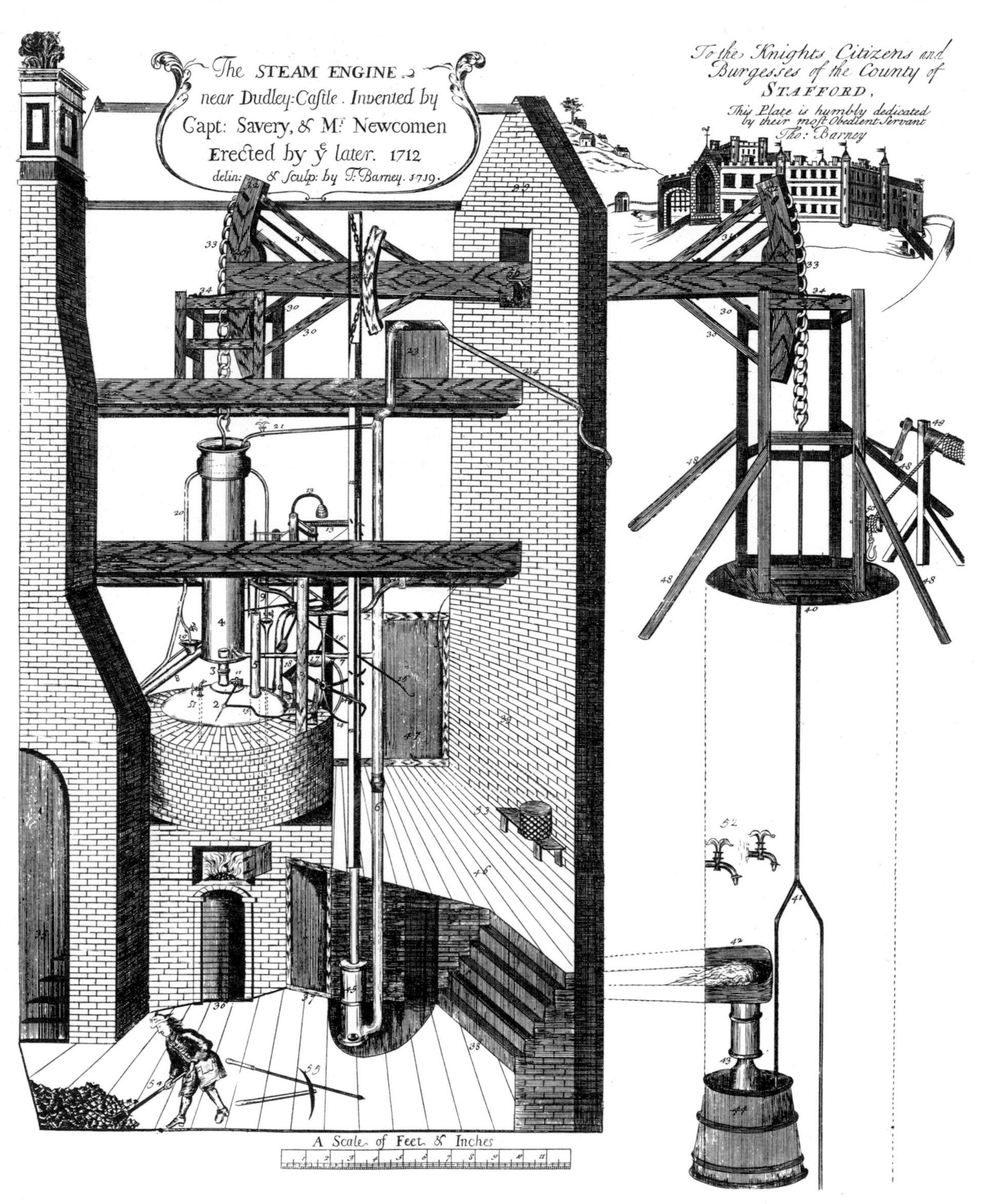
The STEAM ENGINE near Dudley-Castle. Invented by Capt: Savery, & Mr. Newcomen Erected by ye later. 1712
delin: & Sculp: by T: Barney. 1719.
To the Knights Citizens and Burgesses of the County of STAFFORD,
This Plate is humbly dedicated by their most Obedient Servant
Tho: Barney
A Scale of Feet & Inches

water in an hour's time. I saw it going at a distance being come to Birmingham.'

The earliest evidence is, in fact, a print executed by Thomas Barney and published in Birmingham in 1719. This is inscribed 'The Steam Engine near Dudley Castle. Invented by Capt. Savery and Mr. Newcomen. Erected by ye later 1712.' This engraving includes a key to all the parts of the engine and a small picture of Dudley Castle. Although this appears to be only a decorative embellishment, it may be significant that it shows the façade of the castle as it would have appeared when viewed from the Tipton side before the fire which seriously damaged it in 1750.

On one surviving copy of the engraving, a note has been added in contemporary handwriting. This reads: 'Vibrates 12 times in a minute and each stroke lifts up 10 Gall of water, 51 yards p'pender', which represents a horsepower of about 5½. This statement and all other dimensions given agree exactly with the figures stated by Triewald.

Uncertainty over the exact site of the engine prompted a paper to the Newcomen Society by Dr T. E. Lones in 1932 in which he presented the case for a 9 acre site at Tipton which is well within sight of the prominent landmark of Dudley Castle. His conclusion was based on a study of the coal measures in the area and in the light of the known depth of 51yd from which the engine pumped. The area suggested is the site of early coal mining activity and is part of the large and ancient Coneygree Park, for centuries the hunting ground of the Lords of Dudley. This park contained the outcrop of the famous Dudley Ten Yard coal seam on which the great prosperity of the industrial Midlands was to be based.

More recently, documents have been discovered in a private collection which refer to the work of James Shaw, an attorney of Dudley, who became involved in the affairs of the Lords of Dudley. In particular, he was engaged to settle affairs between Ferdinando Dudley and John Lord Ward who both inherited interests in the estate of William, Lord Dudley and Ward on his death in 1740 at which time the Barony of Dudley and Ward separated. Shaw's work involved the valuation of various items of mining equipment, including a fire engine built at the Dudley Wood Colliery in 1736. During this work he found that a cylinder had been bought from Coalbrookdale in 1736 for what was known as the 'Old Engine' at the Coneygree Coalworks, Tipton. The affairs between Ferdinando Dudley and Lord Ward developed into a Chancery case and fortunately, the documents and depositions of the witnesses remain and contain a great deal of evidence upon this engine and the operation of the various coalworks of the area.

The most significant are depositions by Daniel Hawthorne of Walsall in 1748 when he was aged 60 and by John Caddick in 1754 when he was aged 73. Caddick stated that there were then three engines on the Dudley's estates, one at Tipton, one at Dudley Wood and one at the Parke, and they were erected 'about thirty years ago' giving an approximate date of 1724. Fortunately, Hawthorne's statements are much more positive and he gives the history of two fire engines at Tipton, in addition to one at the Parke and one at Dudley Wood. The first of the engines at Tipton had been removed to one of Lord Ward's collieries at Broadwaters, near Wednesbury, and this had a cylinder of 22in diameter, working pieces, buckets and clacks (valves) all of brass and cost when new £450. The other engine at Tipton was called the 'New Engine'.

He then refers to the other engines and gave his opinion that they would if sold be each worth about half their new value except the engine at Tipton called the 'Old Engine' which, 'being the third that was built in England was set up at more expense than such a one might then be built' and so it would be worth only £150. He had made it his chief business for 30 years to set up such fire engines and this confirmed the authority of his statements. It thus becomes clear that this 'Old Engine' is in fact the 'Dudley Castle' engine and the statement that it was the third in England adds weight to the claim for earlier engines, albeit less successful, in Cornwall. Owen Bowen of Dudley, a bailiff of coal pits, also knew of the engine, having by 1754 been employed in coal pits for 20 years and confirmed that the original brass cylinder had been taken out and replaced with a cast metal (iron) one of larger size. Details of the new cylinder are given in the Coalbrookdale accounts being invoiced on 29 October 1736, as '1 cillinder and bottom at 25C – 3Q – 2Lbs at 32/6d. cwt = £41. 17s. 5¼d.' The old cylinder was sold by James Shaw in 1742 for scrap. Another witness told of a new boiler being supplied and fitted which cost about £25. The size of 22in diameter quoted by Hawthorne would be that of the new iron cylinder and the original brass one would be 21in diameter, as recorded by Barney and Triewald.

Subsequent to 1740 the engine was removed to the Level Coal Work, and before 1752 the engine except certain parts was removed to Willingsworth Coal Works, near to the site of an earlier attempt by Thomas Savery to drain a pond, Broadwaters, with one of his engines. According to Dr Wilkes the Savery machine had been 'laid aside as useless'. The canal act of 1783 authorising the building of the branch of the Birmingham Canal Navigation into Wednesbury, refers to its proposed termination 'near an old engine at Broadwaters, the property of the said Lord Viscount Dudley & Ward' and it is possible that this engine was in fact the re-erected 1712 machine. It is shown on an original canal survey of 1785 at the extremity of the Broadwaters Branch a few yards from the Broadwaters.

Various mining leases remain in the Dudley archives and these show that at the time of the engine building one of the main areas being worked at Coneygree was that known as Lady Meadow. From early plans, this large meadow is shown to be near the intersection of the present canal and railway at Coneygree, Tipton. From this position, the ruins of Dudley Castle stand out on the skyline and their elevation is exactly that shown on the Barney engraving. It would seem, therefore, that the engine was built in this area and that the inclusion of the vignette had particular relevance.

Indeed a plan dated 1794 by Richard Court shows an area immediately adjacent, which may well have been included in the ancient Lady Meadow, as 'Engine Piece', and this must be further confirmation.

It seems clear that the engine arrived in the Midlands directly as a result of the Baptist connection with Bromsgrove which has already been mentioned. The Baptist church there was joined by Humphrey Potter, a mercer and Overseer of the Poor, about 1683, and he became deacon in 1689. About 1700, Potter registered his house as a meeting house and at about the same time, built the first meeting house in Bromsgrove. When he died in 1719, he granted the new meeting house to its trustees provided it was still used for worship and they were to take care of it and maintain it in good repair. The first trustee named was Mr Thomas Newcomen of Dartmouth and others were from Bromsgrove, Pershore and Bewdley. For how long Newcomen had been closely connected with Bromsgrove is uncertain, but in view of his teaching at Dartmouth by John Flavell from Bromsgrove it

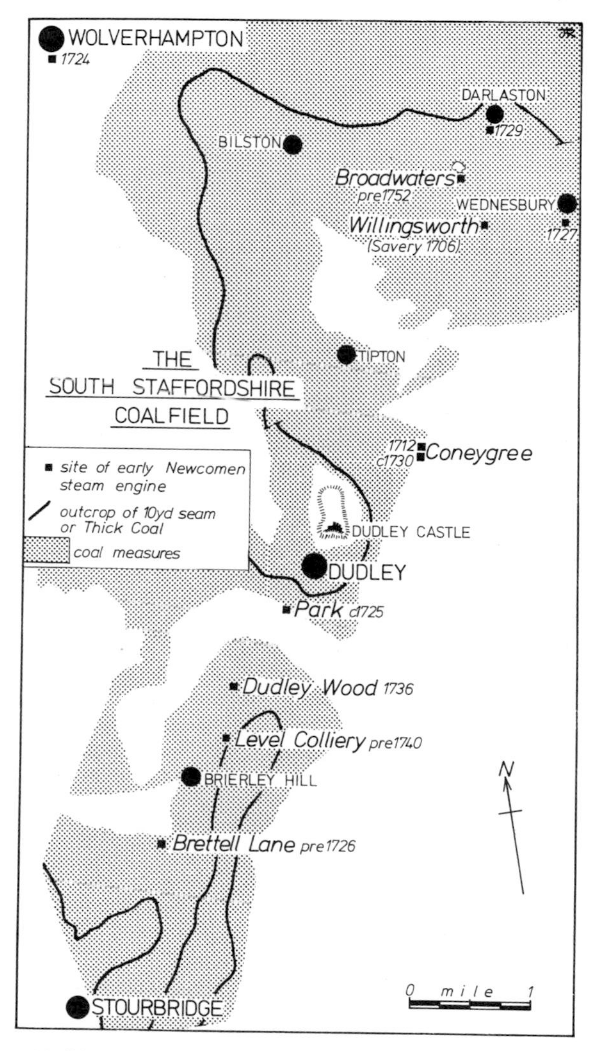

20 Plan showing the approximate sites of the early Newcomen engines in the South Staffordshire Coalfield.

may have been from his early days. It is quite logical that when he visited the Worcestershire forges, he would stay with Baptist friends, and would become familiar with the problems of the Midlands coal industry and the difficulties with drainage at the ever increasing depths at which they had to work.

Bromsgrove had strong religious connections with Dudley, where a Baptist church was flourishing at Netherton, about a mile from the Castle. One of the leading citizens of Bromsgrove was Josias Bate who had a son Josias of Dudley, who died in 1717, leaving his young son and daughter in the care of their grandfather at Bromsgrove. Furthermore, the grandson, also Josias, was to inherit when he became 21 a share of the 'Cunnagree Coalworks' in the Parish of Tipton, which share had been purchased from the 'Trustees and guardians of the Lord Dudley' at a time between 1701 and 1717. Here then is the direct connection from Dartmouth through to Coneygree and the Lords of Dudley for whom the engine was most probably built at the direct charge of Newcomen, who would receive a fee to drain the mines as he did with other early engines. Doubtless, as in other cases, the Lords of Dudley would later buy out the rights to operate the engine themselves and to set up other engines for premiums agreed with the Proprietors who operated the patent following Savery's death in 1715.

The records of the Netherton Cinder Bank Church have survived and contain some surprising references. A list of members is recorded to which is added the names of an Elias Newcomen and a John Dunford. These two members came from elsewhere and were noted as being 'received by a letter about 27 twelfth month 1710', ie 27 February 1711, and a later note adds that they had 'Gone from us'.

The will of Humphrey Potter in 1719 was witnessed by Elias Newcomen, John Dunford and Ezekiel Trengove. It is clear that Newcomen and Dunford arrived at Dudley to build the engine and Trengove, who was associated with the Proprietors of the patent, was almost certainly a relative of Charles Trengove of Dartmouth whom Thomas Newcomen claimed as a relative. This Elias Newcomen was most unlikely to have been Newcomen's son since he could not have been more than 5 or 6 years old in 1711. Newcomen had a nephew also named Elias who was certainly concerned in building other engines at Winster in Derbyshire in 1719 and in 1721. His family were Baptists and came from Chard in Somerset, his father being John Newcomen, an apothecary, and his mother Margaret who died in 1701. Elias had at least two infant brothers who died in 1697 and 1698 and he may therefore have been aged 17 or more in 1711. He was to die about 1725/6 after catching smallpox in 1723 in Spain which left him in a weak condition.

It is now appropriate to refer again to the story by Beighton and Desaguliers that the engine was built for a Mr Back of Wolverhampton, the only such reference. Earlier, given the lack of other evidence, this statement led to speculation and Dr W. O. Henderson in 1948 prepared a paper supporting a site adjoining the Wolverhampton-Walsall road, where there had been early coal workings.

He identified the Mr Back with a Mr William Bache of Wolverhampton who died in August 1712. Much more evidence has now been discovered on this man who certainly was managing mines 'in or near Wolverhampton' for 3 or 4 years before his death and these had been run at a loss which amounted to over £1,000 at his death, 'on account of the charge in getting the coal and erecting of engines for the drawing of water out of the mines'. There is no indication that this reference is to anything but normal pumping equipment and many deponants in a Chancery case concerning his affairs all speak of intimate knowledge of this man, but make no reference to a fire engine which at the moment of his death would have been unique in the world. It is unfortunate that we do not yet know the location of Bache's colliery reported as 'in or near Wolverhampton'. Should this have been in part of the Coneygree then the reference would not conflict with other evidence.

The other reference to Wolverhampton as the site comes from the *History and Antiquities of Staffordshire* (1798-1801) by Stebbing Shaw written almost a century after the events, and apparently quoting notes of Dr Wilkes (1690-1760). These notes have been searched by N. W. Tildesley of Willenhall, but do not now contain the statements quoted so that the case is much

21 'Atmospheric engine, Griff, Warwickshire', from Desaguliers' *Experimental Philosophy,* 1744. This drawing is similar to that by Henry Beighton in 1717. The cylinder was about 16in diameter by 9ft long.

weakened. In spite of the considerable volume of evidence produced in recent years from which the introduction of the engine has become better understood, there is no contemporary reference to the building of an engine at Wolverhampton until 1724.

A reference from Sweden has finally clinched the matter. Jonas Alstromer, a Swede, visited England in 1720 and visited many engineering centres, writing down his discoveries in great detail. Coming to Wolverhampton he met Thomas Barney and discussed the local iron and mining industry. He then recorded in his diary: 'A little way from this place there is a fire-pump, the first machine built in England, which draws up water from some coal mines . . . these installations are worth seeing. Thomas Barney, file manufacturer, has engraved the fire engine on a copper plate and has printed it together with a description, on two sheets of paper that together cost 2s.'

This is further confirmation that this was the first engine and Barney clearly shows it to have been near Dudley Castle. Thus the recently

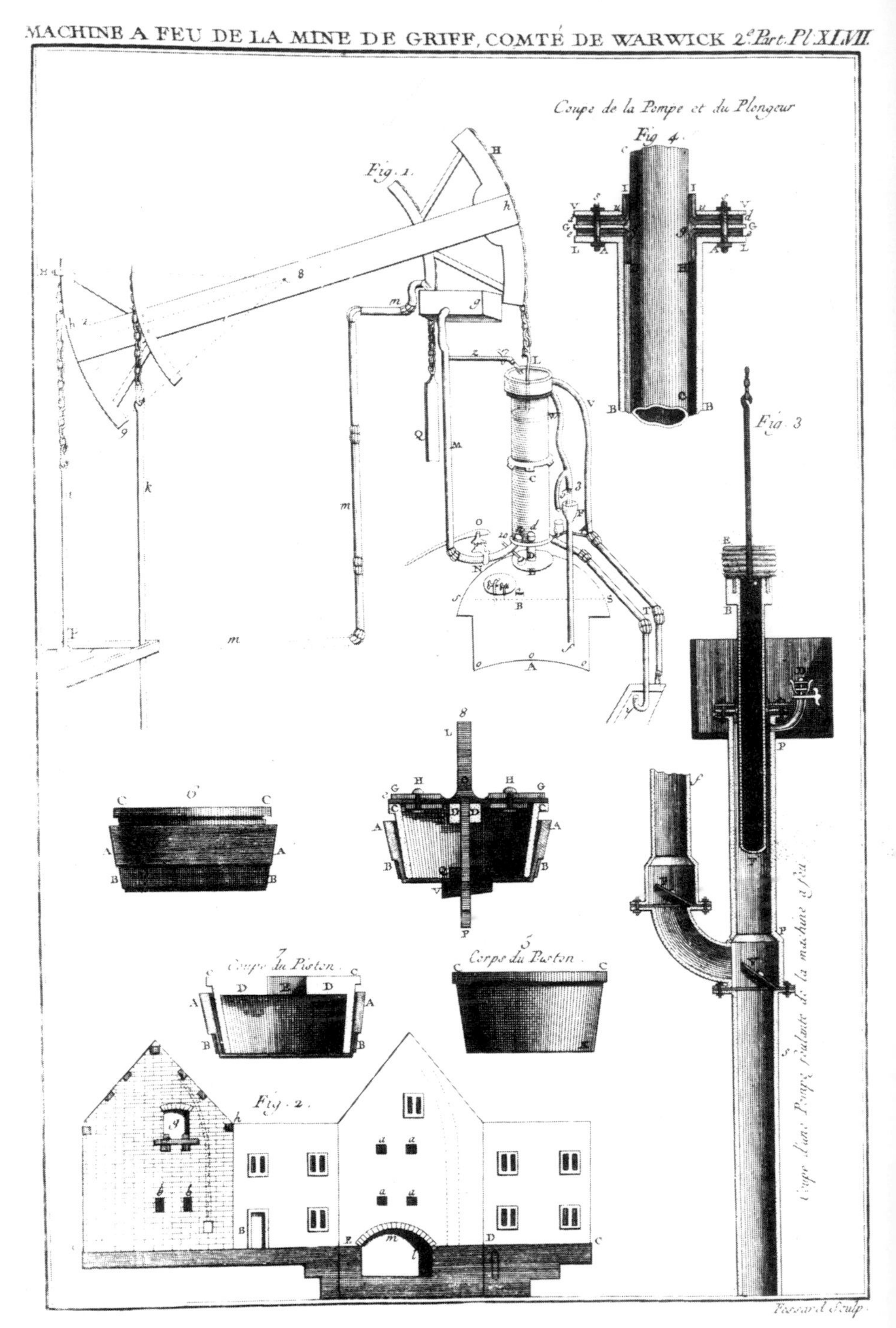

22 A plate showing details of the engine house, piston and pumpwork at Griff. From J. C. Morand, *L'Art d'Exploiter les Mines de Charbon* 1768. The details are similar to those indicated by Desaguliers.

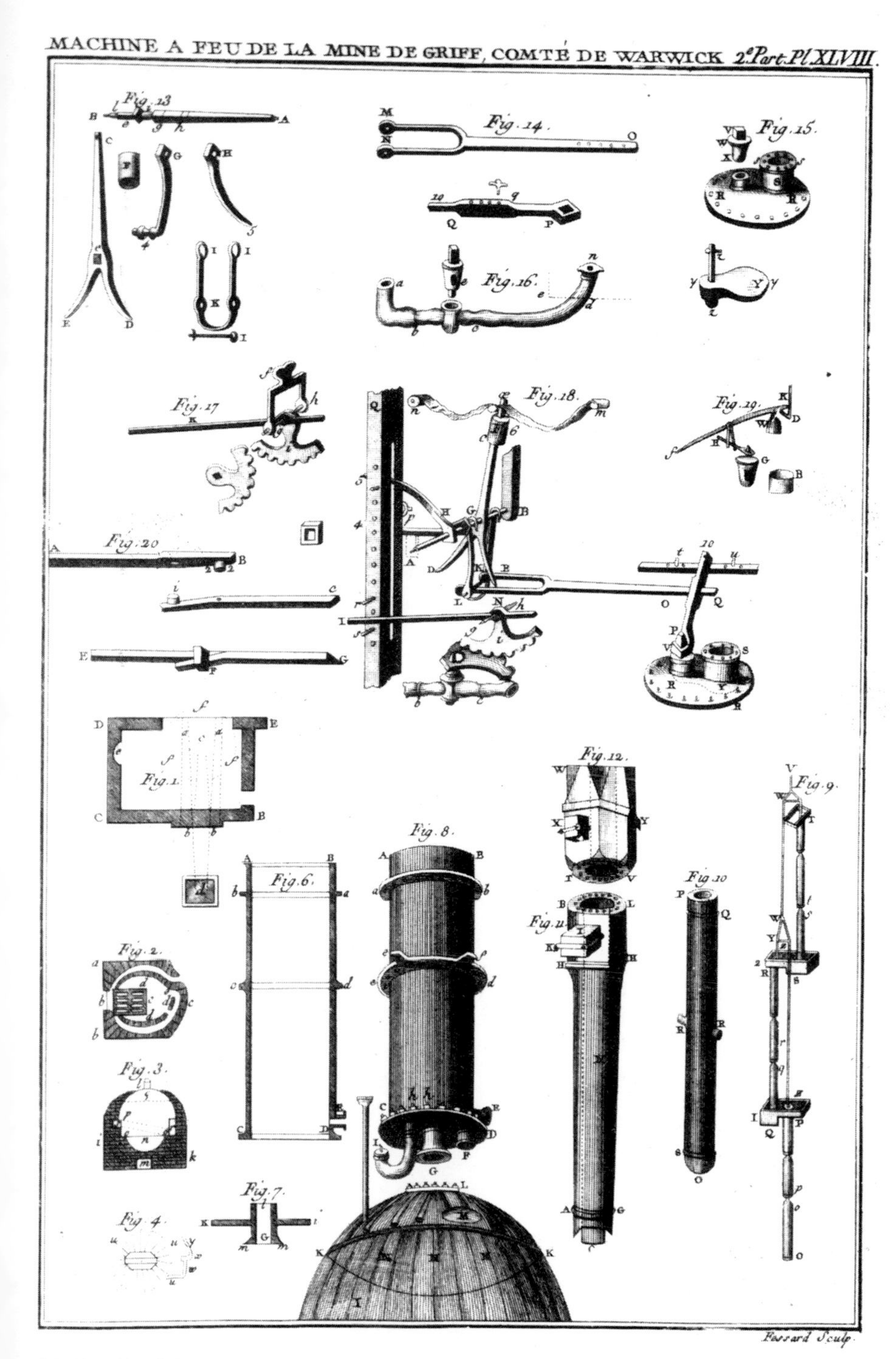

23 Details of the boiler, cylinder, valvework and pump trees of the Griff engine. From J. C. Morand.

discovered evidence has established beyond all doubt that a very early engine was built on the Coneygree Coalworks at Tipton near to the Dudley border and overlooked by the castle. This siting is supported by all other contemporary evidence — the Barney print, Triewald and Kelsall, and geologically fits the area determined as most likely. Further, Elias Newcomen and John Dunford came to Dudley for a period at a time exactly coinciding with the building of the engine which would probably take at least twelve months or more to build during which they were to make 'a great many laborious attempts' before, as Desaguliers states, they were able to make the engine work.

It is possible – indeed probable – that in the seven years which elapsed between the completion of the engine and the making of Barney's print, improvements were made in the light of the experience gained with this and the other engines which followed. Yet even when this has been allowed we can but marvel at the genius of its creator. The engine Barney illustrated so clearly and with only a few minor errors of detail, represents a stupendous advance upon anything that had gone before. In its purposeful and ordered complexity it makes Savery's crude pump or the scientific toys of Papin and his fellows seem so primitive by comparison that it is scarcely credible that so few years separate them. The wonder is not that Newcomen spent anything from ten to fourteen years on his invention before he achieved success, but that such a staggering advance could have been made by one man in a lifetime. The print shows the self-acting valve gear and indeed every feature which, with refinements of detail only, would characterize the steam engine down to the days of James Watt and stubbornly persist for many years after. Seldom in the history of technology has so momentous an invention been developed by one man so rapidly to so definite a form. When, in addition, we remind ourselves of the relatively simple means at Newcomen's disposal in 1712, then we can scarcely fail to regard his achievement with a wonder akin to awe.

The next engine built by Newcomen at Griff, near Nuneaton, Warwickshire, is more positively identified as to time and location for there was an agreement dated 27 April 1714 by Thomas Newcomen in which he agreed 'at his own charge as soon as conveniently might be to set up on some part of the coalworks called Griffe an engine to draw water by the impellant force of fire which should draw or cast up 70 Hogs-Head of water per hour, if so much water be there, not above 47 yards from the sough or level, at and for the weekly sum of £7, payable on Saturday of each week.' The engine consisted of a copper boiler, a brass steam-barrel and piston, two pit barrels of pott metal (iron) and other pipes and cisterns and was built soon after the date of the agreement, the cylinder being about 16in diameter by 9ft long. Doubtless Newcomen would bring to bear all the experience gained from the Coneygree (Dudley Castle) engine and it seems to have represented a definite stage in the development of the engine for at least two later agreements in 1717 and 1718, referred to engines being built 'according to the method now in use at Griff'.

The first proposals for an engine at Griff are noted by Beighton and Desaguliers without substantiation, as being in 1711, at which time the colliery was in the hands of Sir Richard Newdigate of Arbury, Warwickshire. In 1713, he leased areas of his estates to a partnership of Richard Parrott, his eldest son Stonier Parrott and George Sparrow and this partnership was to have a profound effect upon the introduction of the engine. The Parrott's were of Bignall Hill, Audley, Staffordshire and their partner Sparrow was of Chesterton, Wolstanton, also in North Staffordshire. By the beginning of the eighteenth century they were already deeply involved in a variety of enterprises such as coal and ironstone mines, forges, ironworks, brine pits and saltworks in the counties of Stafford, Warwick, Chester, Worcester and Flint.

The engine at Griff has been thought to be that shown on an engraving of 1717 by Henry Beighton, but it now seems that this is unlikely and it is more certainly illustrated by a plate in Desagulier's book which is different in the detail of the valve gear. The engine shown by Beighton is most likely that he built himself at Washington, Newcastle on Tyne in 1717.

The next engine was agreed some time between April 1714 and November 1715 and was built at Woods Mine, Hawarden, Flintshire. The lands had been leased for a period of 41 years from

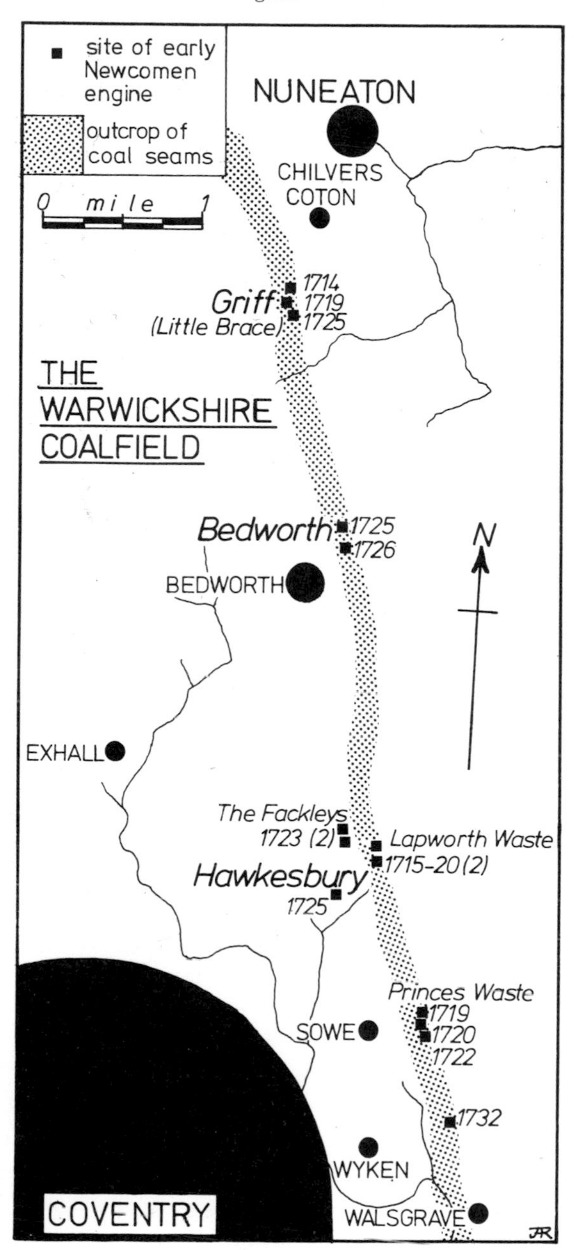

23a Location of early Newcomen engines on the Warwickshire coalfield.

Thomas Crachley of Broadane, Flintshire, by William Probert of Hanmer, Flint, and Thomas and Robert Beach of Meaford, Staffordshire. Thomas Newcomen had made an agreement with these lessees, together with a Thomas Brewer, in which he was entitled to a third share of the profits of the mine. The engine and other preparations were reported to have cost over £1,000.

At about the same time, during 1714/15, another engine was being built at Moor Hall, Austhorpe, Leeds. This engine does not appear to have been particularly successful and went out of use after four years. The engine, however, is important for two reasons. It appears to have been built by Newcomen's partner John Calley, and Calley's son, who were both also involved at Whitehaven where another engine was being built. There were difficulties during September 1717 at Austhorpe and John Calley, Sen., went over to help with the engine. He was taken ill on 19 December and a letter arrived at Whitehaven on Christmas Day 1717 telling young Calley that his father was so ill that there was no hope for his life. Calley set out for Leeds the following day and may have arrived just before his father's death, for he was buried on 29 December 1717.

Quite as importantly the site of this engine was near to where John Smeaton was born in 1724, but the local memory of the engine remained to quicken that interest in the steam engine which Smeaton later turned to good account. He made enquiries about the engine when he grew up and was told by an old man who had worked it that the cylinder was 23in in diameter and 6ft stroke and that it pumped at the rate of twelve strokes in a minute when self acting, but could be worked up to fifteen strokes a minute by the handgear. The depth of the pit was 47yd and the pumps, which were of 9in bore and arranged in two lifts, raised the water 37yd to an adit.

The engine at the Ginns Pit, Whitehaven, for which the agreement was signed on 10 November 1715 is now the most well-documented of the early engines. It was built for Mr James Lowther by 'Thomas Newcomen, Ironmonger of Dartmouth, Thomas Ayres and John Meres' and it was to have a cylinder at least 16in diameter within and 8ft in length at a yearly rent of £182. Its cylinder as built was in fact 17in diameter. Thomas Ayres was a mathematics and navigation teacher who became a Fellow of the Royal Society in 1707. The mention of Meres acting with Newcomen before the formation of the Proprietors is most interesting. The Lowthers' mining agents were the brothers John and Carlisle Spedding and many of their papers remain; in particular detailed reports prepared by John, together with his letter books in which are copied his frequent letters to James Lowther who spent much of his time in London. These tell the story of the abandonment of much of the mines by 1705 due to the inability of old water gins and a wind-driven pump, set too low in the valley to be effective, to cope with the influx of water at the depths they were having to work.

In October 1712 Spedding writing to Lowther, referred to Captain Savery's invention which was the only thing considered to be of any service to them and which Lowther could judge after he had 'seen some experiment of it'. Here he must have been using Savery's name as the original holder of the patent just as the 1719 Barney engraving gives the inventors as 'Capt. Savery and Mr. Newcomen'. As mine owners, they would be fully aware of the original Savery engine first publicised in 1698 and illustrated in print in 1701/2 and well before 1712 known to be unsuitable for the heavy duty of mine drainage. By the end of November, Lowther had reported great satisfaction in what he had seen and Spedding hoped the invention would be useful in several other places as well as the drowned Prior Band at Whitehaven.

It is interesting to speculate on what Lowther saw. The wording certainly indicates the demonstration of a model such as that described by Triewald. However the word 'experiment' at that time also meant 'trial' and it may be that a visit was paid to the 'Dudley Castle' engine then working at Tipton, which would support Triewald's account of visitors coming to the site.

It was to be another three years before Lowther agreed for the engine and the story can be taken up again in August 1715 when Lowther held discussions with Newcomen, presumably in London. Various calculations were made on different systems of draining the mine, but Spedding would not advise Lowther on any system unless he could agree for the engine.

Newcomen's proposals arrived and showed a considerable saving over the use of horses. His scheme was to pump 23yd perpendicular and 33yd on the slope of the dipping coal seam. Newcomen's articles arrived during October and were considered to be drawn 'very clear in every particular' with nothing needing explanation or liable to lead to disputes afterwards. After the agreement was signed, a drawing of the engine house arrived and preparation went ahead. The boiler was made in Newcastle and brought to Whitehaven on waggons which had to be specially widened to take the great load. The other parts arrived by ship from Bristol via Dublin since Whitehaven had no direct service with Bristol at the time.

The engine was built and tested over the winter of 1716/7 but great difficulty arose with the timber pump trees and John Calley was kept very busy trying to stop the leaks. Eventually Lowther asked Newcomen about the price of suitable trees from the West of England and was quoted 12d a foot. Work continued through weather 'with snow thicker than it had been known in the memory of man', and by the end of March they had succeeded in getting the water down to a depth 32yd below the drainage level and had about 4ft further to drain. All was going well and Mr Calley was congratulated upon the work he had done at the pump trees — the water was as low as it was needed and the engine coped very easily with its task.

Unfortunately, this hopeful situation brought about by the introduction and successful work of the engine was shattered when in early April 1717 much of the roof of the workings collapsed suddenly due to the failure of the pillars weakened by their long spell under water. The brass pump barrel at the bottom of the pit was buried in rubbish and other feeds of water entered the workings. They were not able to start the engine again until 11 April and then got down the water sufficiently to enable them to start lowering the main shaft another 9yd and drive a new drainage level.

Difficulties with the boiler were to follow and the quality of the water was blamed. It corroded iron very quickly, but had less effect upon lead. Copper patches were added to the area near the surface of the water, but were eaten away again in a fortnight and the boiler leaked so much it put out the fire! They then lined it around with sheets of lead held together with lead nails which served until a new iron boiler recommended by Newcomen arrived by sea from Frodsham, having been supplied by Stonier Parrott from Bignall Hill. It was wider at the bottom and narrower at the top than the first boiler and had a top of lead. Desaguliers claimed that Parrott invented the iron boiler and there are a number of contemporary references to his supplying boilers at very early dates.

By Christmas 1718 this boiler had failed from external corrosion of the iron due to a continual leak between the lead top and the iron sides. The old one was again patched up and put back into service. In spite of all these difficulties Spedding was still able to confirm that in his opinion there was nothing that would do their business so well and be less liable to accidents than the engine and if Lowther could purchase the lease he was persuaded 'tis the cheapest, safest, and the best way of keeping the colliery dry'.

Richard Bradley writing in 1718 ranked this Whitehaven engine 'above all others that I have yet heard of for ingenious contrivance'. Dr Stukeley visited the Ginns Pit in 1725 where he saw the engine working at fourteen strokes a minute lifting 140 hogsheads in an hour with moderate working.

It has been thought that there was an early engine in the north of the country, the information coming from a map of Tanfield Lea Colliery, near Newcastle-upon-Tyne. The oringinal survey map was made by William Cuthbertson for Gilbert Spearman, the estate owner, in October 1715 and shows an unmistakable representation of a Newcomen engine together with the site of an 'intended engine'. There is a verbal tradition, that this engine was erected in 1714, but recent authoritative examination of the map has concluded that it is a late eighteenth-century copy and thus the engine references would be added at a date later than the original survey. There is no other contemporary evidence and various early 'views' of the colliery by mining engineers make no reference to the engine. George Liddell, writing to John Meres, stated that his company were the first to agree for an engine in those parts and that was for an engine at the Park on 15 May 1715. The early dating of the engines at Tanfield Lea must therefore be discounted.

Another early engine may possibly have been built by Stonier Parrott as a result of a proposal to pump water at Broseley, Shropshire, in May 1715 from a depth of 47 yards with a cylinder 16in diameter by 8ft long at a rent of £20 per year, but no confirmation of its building has been found.

So far as it is known, this concludes the list of Newcomen engines which were either at work or had been contracted for by the time of Savery's death in 1715. His death marks an important turning point in the story of Newcomen because it brought to an end the original partnership. With engines at work in four or five English counties and in Wales, Newcomen had made a remarkable beginning. The long years of experiment, the demonstrations, the journeyings to distant parts in all weathers and the successful building of the fire engines had ended in great success. The power of the atmosphere had been harnessed through the agents of fire and water and the demands of the world waited to be satisfied. Clearly the capacity of the members of the Newcomen and Calley families to cope with the countrywide demand for their services was becoming stretched. A larger organisation and the leasing of rights was necessary to satisfy the demand.

CHAPTER 4

The Proprietors of the Engine Patent

The Livery Hall of the Worshipful Society of Apothecaries of London remains today largely as it was re-built after the Great Fire of London in 1666. Except for a break of three years from 1685, John Meres, Sen, was clerk to the society from 1672 until his death in 1691. John Meres, Jun, succeeded his father on 30 June 1691 following legal training and was to hold the office for 35 years. At the onset of the war of Spanish Succession in 1702 the society obtained a monopoly right to supply medicines for the fleet. Meres drew up the articles of this company known as the Navy Stock Company, and was in effect, its first secretary and treasurer.

It is probable that through the activity of this company that Meres came to know Thomas Savery who took up his post at the Commission for Sick and Wounded in 1705 and the two organisations had many occasions for direct contact on matters of common concern. Savery joined the Royal Society in 1705, and at about the same time persuaded the York Buildings Company to invest in one of his engines. The success of the Navy Stock Company led the Apothecaries to reorganise their own laboratory and in 1713 Meres was involved in the setting up of the Laboratory Stock Company.

Following the end of the war Savery was dismissed from his duties in 1713. He died shortly afterwards on 15 May 1715, having made a will on his deathbed which bequeathed all his estates and rights in his inventions to his wife Martha. The will, probably drawn up by John Meres, was witnessed by John Meres; Savery's clerk, Walter Davis; and Catherine Alexander, his sister-in-law. Savery's widow was left in a difficult position and would have been responsible for his Admiralty debts, so Meres devised a plan which would safeguard her and also permit the more rapid exploitation of the new source of power. Meres arranged for an annuity to be secured upon his property in Paternoster Row, in return for which she would make over her rights in the patent. He then launched a joint stock company, similar in constitution to the Navy and Laboratory Stock Companies known as 'The Proprietors of the Invention for raising water by fire'. The management was in the hands of a committee of their members elected by the stockholders known as 'The Committee appointed and authorised by the Proprietors of the Invention for raising water by fire' to make decisions, and Meres himself was secretary and treasurer.

It seems that the company was launched by public subscription since there appears to be no linking connection between the members and they were not connected with the Apothecaries.

An entry in the cash book of Sir James Lowther of Whitehaven gives us a clue to the details of the company. It reads as follows:

> Fire Engine. 10 May 1721. Then bought of Mr. Edward Burt of the Admiralty Office one share (there being eighty in all) in the Engine for raising water by fire, there is a term yet to come upon the patent and Act of Parliament of about 13 years. I had a transfer of the said share made to me in the transfer book kept at Mr. Meres his office at Apothecaries Hall in Blackfriars which is all I have to show for it except Mr. Burt's receipt for £270. They propose to make a dividend of £30 a year. [Burt was Chief Clerk of the Navy.]

The total capital was therefore £21,600 and the proposed dividend represented just over 11 per cent. It is not clear whether holdings were limited to one share per member as is understood to be the case in Meres's Navy and Laboratory Stock Companies. Lowther's first four dividends were in fact only £20, but this increased to £30 and in 1728 and 1730 the dividend was £40. After the expiry of the patent the dividend continued for a further four years to 1737 but fell to £5, with a final payment of £10.

It seems possible that Humphrey Potter also held one of the shares for his will in 1719 refers to a bequest to his grand-daughter of 'what Interests I have in the fund established on the fire engine and my share thereof'.

The first transaction of the Proprietors was most important and led to the building of a number of engines for which they were to receive royalty payments at no cost to themselves. This was occasioned by the success of the engine at the Griff Colliery which had been worked by Richard and Stonier Parrott and George Sparrow since 1713 and for whom Thomas Newcomen had agreed to act as drainage engineer in 1714. The mining partners wished not only to take over the maintenance of the engine, but to have liberty to build and operate, at their own costs, as many engines as they should need to pump out the water from their various collieries.

The agreement dated 7 March 1715/16 was made for a term of 99 years, the colliery partnership paying £150 for the first half-year and thereafter £420 per year for each mine drained. Should any mine become extremely lucrative and produce over 20,000 stacks of coal per year a further payment of £100 for each 5,000 stacks would be made. Alternatively, the Griff Colliery could be leased by the Proprietors for £300 per year, if the Partnership decided they did not wish to continue working it. The agreement was signed by the colliery partners and by the following on behalf of the Proprietors: Thomas Newcomen of Dartmouth, Devon, Gentleman, John Meres of London, Gentleman, Edward Elliott of London, Gentleman, Thomas Beake of the City of Westminster, Esquire, Henry Robinson, Citizen and Mercer of London, William Perkins of the City of Westminster, tallow chandler.

This list is significant in that it shows that Newcomen was certainly one of the Proprietors from the inception although his name does not appear in any later similar document. We know from affairs at Whitehaven that he was actively involved there in 1716 and 1717 and from later evidence that he was in Cornwall for the whole of 1718. His was undoubtedly not the role of the financier but of the engineer, active in the field where his knowledge and experience could best be utilised.

An advertisement was included in the issue of the *London Gazette* for 11-14 August 1716 which gave the world details of the arrangements which had been concluded. It reads as follows:

> Whereas the Invention for raising Water by the impellant force of Fire, authorised by Parliament, is lately brought to the greatest Perfection; and all sorts of Mines, Etc may be thereby drained, and Water raised to any Height with more Ease and less Charge than by the other Methods hitherto used, as is sufficiently demonstrated by diverse Engines of this Invention now at Work in the several Counties of Stafford, Warwick, Cornwall and Flint. These are therefore, to give Notice, that if any Person shall be desirous to treat with the Proprietors for such Engines, Attendance will be given for that Purpose every Wednesday at the Sword-Blade Coffee-housc in Birchin Lane, London, from 3 to 5 of the clock; and if any Letters be directed thither to be left for Mr. Elliot, the Parties shall receive all fitting Satisfaction and Dispatch.

The Committee of the Proprietors and their Background

Thomas Newcomen and Edward Wallin

Later lists of the committee members omit the name of Thomas Newcomen but include that of Edward Wallin of London, Gentleman. Wallin was in fact a fellow Baptist of Swedish descent and served from 1702 until his death in 1733, as pastor of the chapel at Maze Pond, Southwark. At his funeral service on 18 June 1733 the Rev John Gill described the life of this man who had entered the Ministry when young and had cheerfully led his small flock through difficult times: 'His large knowledge of, and acquaintance with men and things . . . fitted and gave him an uncommon turn for business; and he was well esteemed.' There is no doubt that Wallin acted on the committee in Newcomen's interests and we shall later find evidence of Newcomen staying at his house as a friend in 1727 and he was to die there in 1729.

Edward Elliott of London, Gentleman

It was Elliott who attended the Sword-Blade Coffee House in Birchin Lane, London, between 3 and 5 o'clock to receive enquiries and to act as a contact for the committee. Very little is known of him except that he was a parishioner of St Dunstan's, Stepney and died intestate in August 1717.

Thomas Beake of the City of Westminster, Esquire

Beake was a well known figure in the City, living in Beak Street, Soho, and having the position of under-clerk of the Privy Council and manager of 'The Charitable Corporation for Relief of Industrious Poor'. He was at other times described as King's Messenger, Agent of St Christopher's, and Secretary of the Colony of Maryland from 1714 until his death in 1733. He was buried in St James', Picadilly, his executors being his brother Gregory and a nephew William Sharpe who both, on his behalf, signed the discharge of the agreement in 1735 for Andrew Wauchope's engine at Edmonstone Colliery, Midlothian.

Henry Robinson, Citizen and Mercer of London

Following a seven-year apprenticeship, Robinson joined the Mercer's Company in 1711 and was elevated to the livery on 22 June 1720. He was living in St Margaret's Hill, Southwark in 1748, and died at his home on Wandsworth Hill in 1762, leaving £19,000 and a considerable amount of property. He appears to have been a Baptist and made bequests to Edward Wallin's son, then the pastor at Maze Pond and to the Protestant Dissenting Charity Schools at Horseley Down and Southwark.

William Perkins of the City of Westminster, Tallow Chandler

Perkins was the wealthiest and most influential member of the committee. He became a freeman of the Tallow Chandlers' Company, was elevated to the livery in 1694, to the Court of Assistants in 1709, and later, senior assistant until his death. He was knighted on 20 October 1714 and was friendly with Lord Chancellor King and he may have advised Newcomen to apply to King in connection with a Dutch patent. He died at his estate at Chertsey 5 October 1741 and was buried in the churchyard of St Andrew Undershaft in the City. He left £9,000 to charities and the remainder of his estate to a friend Henry Weston.

Cornelius Dutch, Jun

Dutch was appointed Clerk to the Apothecaries in May/June 1726 following the death of John

24 Cornelius Dutch, Jnr. From a portrait at Apothecaries Hall, London. A member of the committee of the Proprietors who took over the secretaryship from John Meres in 1726.

Meres, for whom he had acted as clerk. He remained clerk to the society for 30 years and a portrait of him, still hangs in their hall. He was a principal beneficiary in Meres's will and took over his duties with the Proprietors, becoming central correspondent for the engine business. He died in 1756.

Ezekial Trengove

A number of the Proprietors' documents and agreements were signed as witness by Trengove who lived or carried on business at Apothecaries' Hall. He may well have been related to Charles Trengove of Dartmouth whom Newcomen claimed as a relative. It is significant that he signed the will of Humphrey Potter of Bromsgrove in 1719. He, like Wallin, must have had Newcomen's interests at heart as a fellow Baptist.

Developments under the Proprietors

In order to encourage the building of engines in the North East, the Proprietors appointed John Potter as their agent and they inserted an advertisement in the *Newcastle Courant* on 27 January 1724 as follows:

> This is to give notice to all gentlemen and others, who have occasion for the fire engine, or engines for drawing of water from the collieries, etc, to apply to Mr. John Potter in Chester-le-Street, who is empowered by the Proprietors of the said fire engines to treat about the same.

John Potter and his brother Abraham were relatives of Humphrey Potter and other members of the same family were also involved in engine building. The will of Humphrey enables the relationship to be determined and other information is provided by the Bromsgrove and Netherton Baptist Registers and the Bromsgrove Parish Register.

Another Bromsgrove Baptist was Joseph Hornblower, who, according to his descendant, Cyrus Redding, met Newcomen when he visited the town to see Potter. Hornblower, too, threw in his lot with Newcomen. He apparently worked for a time in Staffordshire before he went to Cornwall in 1725 to erect an engine at Wheal Rose, on the North Downs near Truro. This, if we accept that there was an engine at Wheal Vor, was the third Cornish Newcomen to be built. Hornblower settled in Cornwall where his son Jonathan became one of the most celebrated engine erectors, while his two grandsons, Jonathan and Jabez, both challenged the Watt patent monopoly.

The first engine actually built following the formation of the Company of the Proprietors was at Yatestoop, Winster, Derbyshire, and this was the first lead mine to be served by the engine. It was built by George Sparrow under the terms of his agreement already discussed 'on behalf of himself and of all the Patentees . . . in a late invented engine called a fire engine', and the document was signed on 18 May 1716. He agreed to supply all the materials for one or two engines, except some large elmwood for the pipes. In return, he was to receive a seventh of all the lead ore raised at the mine and was under £100 penalty if the operation of the mine was affected by failure of the engine to be kept at work. He raised the necessary capital to build the engine by selling his share of the mine in forty-eighths and after the building of the first engine demanded money from these shareholders for the construction of a second one.

Elias Newcomen, the nephew, was at Winster in 1719 and was involved in the building of the second engine about 1721. He wished, because of an involvement with a girl, Ann Stone, to move from Winster, but when he wrote to the committee and to Sparrow, for whom he worked in Staffordshire, they persuaded him to stay on until midsummer 1722. He then left and worked in Spain, selling clothing and materials where he caught smallpox which he just survived. He returned to England and died about 1725/6.

It would seem that Stonier Parrott and George Sparrow already had some arrangement with Thomas Newcomen before the agreement with the Proprietors was signed on 7 March 1715/6, for as early as April 1715 they were in correspondence with George Liddell of Ravensworth, Co Durham, with proposals to build engines. On 15 April 1715, one month before Savery's death, an agreement had been made in which they were referred to as the 'Fire Engineers'. The colliery concerned was 'The Park' and the partners were George Liddell and William Cotesworth of Gateshead, and Francis Baker of Birkham, Co Salop. The fire engine was to be capable of raising 200 hogsheads of water an hour from a depth of 50 yards and the annual rent was to be £300. Other agreements were made in an attempt to secure a monopoly of the drainage rights of all except one of the collieries in the Tyne and Wear.

That the 'Fire Engineers' were in effect part of the Company of Proprietors is shown by letters in November 1717 to John Meres stating that Liddell was the first to agree for an engine in the area and requesting an agreement for another engine at Farnacres. In reply, Meres gave some most interesting information: 'We hope we shall have a barrel and boiler for you in a short time, having some coming round from Cornwall.' It could well be that these would be from an early

25 A drawing by Henry Beighton 1717, probably of his engine at Oxclose. Note the difference in the valve gear from the Griff engine.

engine in Cornwall which had gone out of use and certainly adds weight to the suggestion of engines there prior to 1712. In December 1717 Cotesworth saw both Parrott and Meres in London, confirming again the association between Parrott and the Proprietors. A copper boiler was coming from Wales and Parrott confirmed having made a boiler from saltpan plates (iron) which cost only £37 against £150 for one of copper. It will be remembered that Parrott supplied an iron boiler to Whitehaven in January 1718. Parrott took a lease from Ralph Brandling of Middleton, of Felling Grounds next to The Park on 19 July 1717, and this colliery was to be worked in conjunction with The Park. The engine was eventually built at The Park in January 1718/9 in a house to hold two engines, and the second engine was built but never worked. The quantity of water in the mines was too much for the engines and the attempt failed, Parrott claiming that on top of his £2,000 costs, 'his whole substance in the world and ten times more would not have been sufficient to accomplish the task.' The troubles at the mines, as in a number of other cases, led to disputes which went to law.

Another engine was built in the North East by Henry Beighton, a man closely associated with the development of the engine. He was born in 1686 at Griff, near Nuneaton in Warwickshire, but left to work in the North East, later returning to his native place where he was to die in 1743 and was buried at Chilvers Coton. Prior to December 1717 he built an engine at Oxclose Colliery, Washington Fell, and introduced there a modification to the valve gear which will be described in the next chapter. In the same year, Beighton was 'Engineer' to Parrott when Felling Grounds were leased and lived at the Staithman's House there between 1717 and 1721.

Beighton was the first man of science to make a study of the steam engine and in 1717 he calculated data showing the diameter of pump barrel and steam cylinder required in order to pump a given quantity of water from a given depth. In 1721, this information was published in tabular form, of all places, in *The Ladies' Diary*, of which he was editor. In this table the pump diameters range from 4 to 10in and the cylinder diameters from 9 to 40in. The greatest pumping depth tabulated is 100yd and a 6ft working stroke and a speed of sixteen strokes a minute are assumed throughout.

For many years Barney's engraving of 1719 was believed to be the earliest portrayal of a steam engine in the world, but in 1925 another engraving of a Newcomen engine was discovered in the Library of Worcester College, Oxford. This is entitled *The Engine for Raising Water (with a power made) by Fire* and it is inscribed 'H. Beighton, delin. 1717'. This evidently illustrates another engine built after the Dudley Castle machine and there are certain differences of working detail, but internal injection and self-acting valve gear are both evident. Thus, even if we assume that the Dudley Castle engine did not possess these features when it was built, this drawing positively sets the date of their introduction back another two years. Since we now know of Beighton's involvement in the North East it seems most probable that the engine shown in his print is that at Oxclose Colliery.

Beighton was elected a Fellow of the Royal Society in 1720 and some of his scientific papers are in the British Museum. He returned to Warwickshire before 1725 and he and his mother leased certain collieries from Sir Richard Newdigate of Arbury Hall to whom he appears to have acted as a friend and advisor. Beighton was also a surveyor and published in 1725 a map of the County of Warwick on which he indicated that three collieries in the Warwickshire Coalfield were equipped with fire engines. These were at Griff, Hawkesbury and Fackley. The names and coats of arms inscribed along the border of this map include those of 'Stonier Parrott of Fackley, Gent'.

The second engine at Griff foreshadowed by the Parrott-Sparrow agreement was built about 1719 and there were in fact even more engines in use in the Warwickshire coalfield from Chilvers Coton (Nuneaton) in the north to Wyken (Walsgrave) in the south, than might be gathered from Beighton's map. They were all built by Parrott and Sparrow either in partnership or as individual ventures. As the size of their undertakings increased, so did their personal difficulties and, as a result of many legal battles, they were to become great enemies. Parrott in particular was in financial difficulty and was declared bankrupt in 1732, but Sparrow managed to disengage himself and remain solvent.

The mass of legal documents left as a result of their activities have provided a new field of information from which the operations at the various coalworks can be studied and the building of engines confirmed. In November 1719 the first of two engines, in a double engine house at Wyken came into use, followed by the second in July 1720. The boilers of both were made from iron plates. In 1722 a third engine was built and Henry Beighton was paid £1 1s 0d in 1723 for drawing two maps of the lands and coalmines at Hawkesbury. Two engines were built between 1715 and 1720 on Lapworth Waste at Hawkesbury and about 1723 two more were built at Pickards Fackley, Hawkesbury, these described as large engines, pumping from a depth of 70yd. About 1726 another engine was built in Cooper's Meadow, Bedworth.

Returning again to the North East, we find an engine was built at Byker about 1717/8 for Richard and Nicholas Ridley of Newcastle-on-Tyne. Although the agreement for the engine was dated 4 November 1718, royalties were to be paid from 24 June 1718 so that it appears that the engine building work started well before the legal documents were completed. The engine was to be built 'for the drawing of water only', according to the method now in use at the Griff near Coventry' and was presumably of only moderate size, the royalty being £420 per year, the same as that for Griff. The Proprietors agreed 'to provide a proper and fit person to direct the building of the house necessary for setting up the engine and other matters thereto belonging, and also for the sufficient instruction and teaching such servants . . . as shall be appointed, in the use and working of the engine' and the Ridleys had to pay this person his wages during the time of his service.

Due to the rapid expansion in demand there is no doubt that Newcomen, Calley and the Proprietors had by this time become somewhat stretched in their ability to serve the various mine owners scattered throughout the country. They had therefore to send to Byker a young man of 16 years of age, Samuel Calley, in all probability the second son of John Calley. Marten Triewald was in London at this time and there he met Nicholas Ridley who was concerned over the tender age of the engineer constructing his engine and Triewald was persuaded to go to Newcastle to watch over the operations and preserve the interests of the Ridleys. It is interesting to note that in his account Triewald repeatedly refers to 'the inventors' whom he defines in other contexts as Newcomen and John Calley, although the agreement was clearly with the Proprietors. This adds weight to the evidence showing that Calley as well as Newcomen was amongst the original Proprietors. The story in Triewald's own words reads as follows:

> In the year 1717 when similar engines already had been built at three different places in England, it happened that Messrs Ridley in Newcastle also had made an agreement allowing them to use fire-engines against a yearly payment of £400. They did this in order to make it possible for them to work a very large and rich pitcoal-mine, located near the town of Newcastle and its river Tyne. This mine had always been much coveted, but was so wet that it, within living memory, had ruined two very wealthy families, who knew no other method of drawing the water from the mine and thus making it possible to work, than the use of horse-driven pumps. At the same time the inventors were busy erecting fire-engines in other places and, therefore, it was not possible for Messrs Ridley to get anybody else to help them but the son of one of them called Samuel Calley, a lively youth, 16 years of age. This Calley although practically reared in a fire-engine was rather young and did not, in spite of all his practical experience, know anything at all about theory.
>
> Mr Nich. Ridley who was then staying in London, was not only worried because of the youthfulness of his engineer, but also feared that his competitors and other owners of coal-mines in the neighbourhood might corrupt the youth, so that he would not serve him faithfully. Mr. Ridley, who had known me from early childhood and moreover was aware of the diligence and zest with which I had been studying natural science and mechanics in London, was then guided by the wonderful providence of God, to conceive the idea of trying to persuade me to assist his young engineer, and to watch him in case he did not service his master honestly and faithfully. Thus he persuaded me to go with him to Newcastle, promising to give me the opportunity of learning how to construct fire-engines, and I for my part promised to serve him loyally against a fair reward.
>
> A few days later we arrived in Newcastle where the construction of the first engine was well under way. According to our agreement I did not allow anybody within the space of a year and a day to discover that I understood anything whatsoever about such an engine, but as soon as I saw it at work, I conceived a more perfect theory of it than the inventors themselves possessed to the very moment of their death. Even the great Newton testified in the presence of the Commission-Secretary Herr Skuten-hielm that such was the case and admitted that he never had been able to get a correct idea of the fire-engine from the inventors, because they always ascribed the power to the steam, which, however, only constitutes the agent by means of which the power is created.
>
> The result was that when Messrs Ridley found it necessary to demand from the inventors a larger engine than the biggest one that so far had been built, which had a cylinder 28 inches in diameter, they flatly refused to construct anything larger and claimed that the engine that Mr Ridley wanted, namely one with a 33 inch cylinder, would be completely impractical, particularly because the boiler for the former hardly could provide enough steam to run it. The cause of this conclusion was the false principles concerning the

steam which the inventors harboured in their minds according to which the steam rises or is generated by the boiling water in proportion to the quantity of water in the boiler. In consequence their boilers were made very high, as demonstrated by the Stafford engine, the boiler of which is higher than its width. It is thus evident that the inventors do not know that the boiler must be given a suitable shape. Neither did they know that the flames should be allowed to play all around the side of the boiler as well as on the bottom — not to mention many other improvements which the fire-engine appeared to require according to a sound theory. As I now offered to construct an engine with a cylinder 33 inches in diameter and 9-feet long, Mr Ridley induced the inventors to grant me permission to construct fire-engines. Young Mr Calley was subsequently taught the theory by me and we then formed a Company together and signed a contract to the effect that we should divide equally between us all the income we might derive from the construction and supervision of fire-engines, which contract still holds good whenever I like to return to England.

The Triewald patent of 1722 permitting him to build engines has been found and clearly shows that its only novelty lies in the claim that the engine is to work by the power of the atmosphere and that Triewald had appreciated, as others were also to do, that the original Savery patent was not strictly correct for the Newcomen engine, which operated on vacuum and not 'the power of fire'. Nevertheless, Triewald was able to obtain Royal Letters Patent for a term of 14 years and made a bargain to take quarter shares of the profits with Nicholas Ridley, Samuel Calley and William Prior. It is strange that Triewald's patent had no effect upon the rights of the Proprietors. However Triewald and Calley built the second engine at Byker whose cylinder was 33in diameter by 9ft long.

Papers in Sweden concerning a later legal action against Triewald contain a copy of a letter of recommendation from Richard Ridley, the father of Nicholas, dated 15 February 1723 addressed to Mr Edward Wite (sic) at Lumley Castle. It refers to a proposal by Lord Scarborough to erect a fire engine and reports of Triewald and Calley: 'none know that sort of business better than they or are more careful, diligent or faithful, they having for many years been in my family service'. Triewald then states that they had in fact built four engines for the Ridleys, but does not refer to the building of the engine for Lord Scarborough which was to be at Sunderland. A cylinder was however supplied to him in 1733 from Coalbrookdale. The location of the two later engines built by Triewald is uncertain, but one was probably at Walker Colliery, Newcastle. Triewald returned to Sweden in 1726 where he built the engine at Dannemora.

According to E. Hughes in his *North Country Life in the Eighteenth Century, the North East 1700-1750,* Samuel Calley was still concerned with engines in 1745 and paid £5 per month 'to keep a fire engine going'. It would seem therefore that after Triewald's departure Calley did not progress further as an engineer.

The Triewald patent was in existence when Stonier Parrott prepared his memoranda in 1725. He too observed that the Newcomen engine did not operate within the terms of the Savery patent as 'it was impelled by the atmosphere only and not the least imitation of Savery's engine'.

The seams of the Tyne coalfield form a vast subterranean basin, the rim of which is in the west and north. From this rim the seams dip to a depth of 1,200ft below sea level before they are overlaid by an area of Permian limestone which extends inland from the sea coast south of the mouth of the Tyne. The relationship between this configuration and the surface contours is such that a great deal of the rim of the coal basin lies far below high ground. Consequently it was only in the vicinity of river valleys, the Wear and Tyne and the latter's tributaries the Ouseburn and the Team, that the seams out-cropped and could be worked by surface digging or shallow bell-pits. By the time with which we are concerned these outcroppings had been worked out and pits had been sunk to depths of from 200 to 300ft. Yet these pits too were confined by the contours to the river valleys where the problem of keeping them free from water was very great. This explains why, as Triewald tells us, Byker Colliery in the Ouseburn Valley had already ruined two owners before Nicholas Ridley took up the challenge and why many other coal owners in the area found themselves in a similar plight.

As Triewald's account also reveals, no sooner was the first engine at work at Byker Colliery than

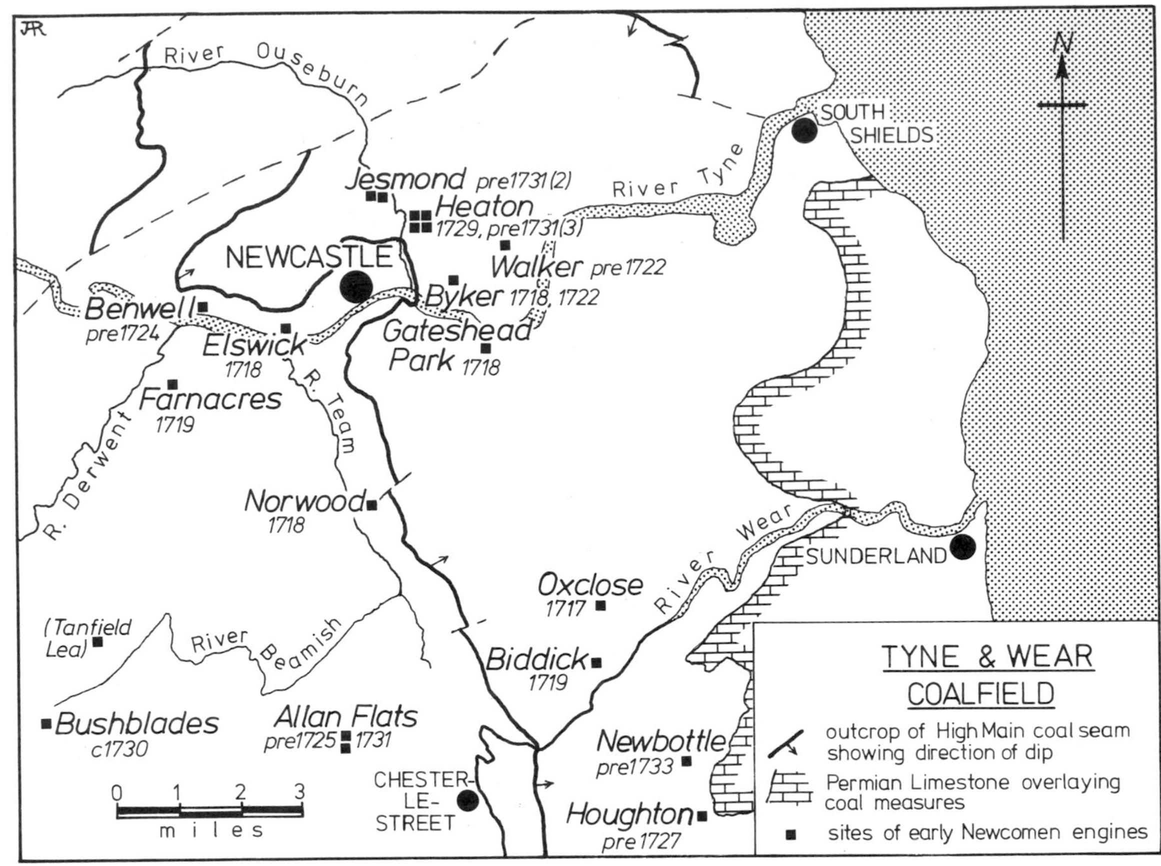

26 Map of the North-East Coalfield showing the sites of early Newcomen engines.

Ridley was demanding a second, larger engine to keep the waters at bay. The ability to work such water-logged levels by the aid of the new power soon led to a still greater influx of water and a demand for more pumping power. It was this vicious circle which made the Tyne Basin so speculative and costly a field for the colliery owner and such a fruitful market for the engine builder. Despite a prodigious expenditure on pumping power, the battle with the waters was ultimately lost, a great part of the Tyne Basin workings being drowned out and not recovered until the late nineteenth century. Developments at other collieries near Byker in the Ouseburn Valley followed the same pattern with the effect that by 1733, when the patent monopoly expired, there were eight Newcomen engines at work in this one small area: two at Byker, three at Heaton Colliery, one at neighbouring Crag Hall, and two at Jesmond. Other engines at work by this date were at Elswick Colliery in the Tyne Valley, at Norwood Colliery near Ravensworth in the Team Valley, at Park Colliery, Gateshead, Farnacres and Biddick Collieries Tyneside and perhaps at Bushblades Colliery which is situated near the headwaters of the Team. To these must be added the engine built by Beighton at Oxclose.

A Swedish traveller Henric Kalmeter (1693-1750) visited the Newcastle area in August 1719 and describes how he saw four fire engines under construction. The first was '1 mile west of the town and on the south side of the river' and this was probably at the Park, Gateshead; the second was the engine at Norwood where he states 'the mine had the inconvenience of water'; the third was at Elswick, and the fourth is a little more difficult to locate. He records that it was at Baikershore probably meaning Byker Shore and it may well have been Triewald and Calley's engine as he records it 'Was now so complete that its so called Boyler & Cylinder were put in'.

Apart from the second Byker engine and the two other Triewald and Calley engines, the fire engines in this Newcastle area brought considerable royalties to the London Proprietors. Triewald's claim that he was a greater master of the principles of the Newcomen engine than Newcomen himself need not be taken seriously. There is no evidence that he contributed any improvement to the engine. It is true that, as he says, the boilers which supplied the early engines were ill-proportioned and inadequate, but here it would seem that he was being wise after the event. The engraving in his book, which purports to show the engine which he constructed at the Dannemora Mines after his return to Sweden, not only portrays no improvements in either engine or boiler, but includes a device which became obsolete as soon as the steaming capacity of boilers was improved.

That Newcomen refused to supply a cylinder larger than 28in diameter for the second Byker engine may be true, but if so his refusal is far more likely to have been due to production difficulties than to any ignorance or prejudice. Nicholas Ridley evidently found a brass foundry in the district which proved capable of casting a 33in cylinder for him. This technical success seems to have made the area self-sufficient where brass cylinders are concerned. Brass cylinders cast in the south of England did not exceed 28in and there was also a great difference in cost. In 1733 £150 was quoted on Tyneside as the price of a 33in cylinder in an estimate for building a new engine at Jesmond Colliery, whereas a 28in or 29in cylinder supplied through the Proprietors for Edmonstone Colliery, Midlothian, cost the owner £250 in 1727.

The very fact that Tyneside led the way by producing larger and cheaper brass cylinders made the area conservative when the Coalbrookdale Company began to offer iron cylinders, which were not only cheaper than brass, but could be cast and bored in the larger sizes which were badly needed. By the time the patent lapsed in 1733, a total of twenty-two iron cylinders had been cast at Coalbrookdale, but of these only one was consigned to Tyneside. All the rest went to the Midlands or to Wales. The exception was ordered by 'Alderman Ridley, Newcastle on Tyne' in 1731. It evidently failed to make a good impression, for twenty years passed before iron cylinders of rapidly increasing size began to supersede brass on Tyneside.

We do not know for certain where any of the early brass cylinders were cast, but as late as 1752 the most celebrated Tyneside engine-wright of the day, William Brown, could write in a letter to his friend Carlisle Spedding of Whitehaven on the subject of cylinders that: 'I have had Several Brass Foundiers at me and as they are so plentiful has some thoughts they will reduce the prices . . .' This suggests that whereas the London proprietors experienced difficulty over the supply of good brass cylinders, a circumstance which favoured their substitution by the Coalbrookdale product, on Tyneside their production was mastered by more than one founder with the result that prices were kept down to an extent which made the iron cylinder less attractive. Brown obtained brasswork for his engines from a founder named John Thompson of Gateshead and it may be that this foundry was producing cylinders at an earlier period. Brass parts were supplied during the early period by William Brookes a brass founder of Bromsgrove and he also supplied at least one cylinder. Newcomen used as agent Nehemiah Champion of Bristol, who was involved with the Bristol brass industry and it seems most probable that some cylinders and other brass parts came from this source.

It is very likely that the brass cylinders supplied by the Proprietors from the south of England were cast in a gun foundry. Although the iron cannon was predominant by this date, Mathew Bagley a bell founder of Chacombe, Northants, was casting brass cannon at the Moorfields Foundry London from 1704 to 1716. When the difficulties and hazards of casting large brass cannon at this period are appreciated the high cost of brass Newcomen cylinders becomes understandable. Charles Ffoulkes in *The Gun-Founders of England* (1937) describes how in May 1716 a terrible explosion occurred at the Moorfields Foundry while a large cannon mould was being filled. It is believed to have been due to the mould being damp. Bagley and fourteen other men were either killed outright or died from their burns and the foundry was closed down forthwith.

The decision to install a Newcomen engine with a 33in brass cylinder involved the Tyneside colliery owner in an outlay of around £1,000 and this would not include the cost of contingent works which might be involved such as the sinking of a new shaft. This was a very large sum in those days and it naturally affected the price of Newcastle coal. In 1738 the London tradesmen, glass-makers, brewers, distillers, bakers, smiths, soap boilers, dyers, lime burners and brick makers, who consumed large quantities of fuel, petitioned the House of Commons in protest against the high price of Newcastle coal. In the enquiry which followed, evidence was given by the colliery viewers which sheds valuable light on the economic background to the introduction of steam pumping on the Tyne. Thus one George Cloughton made the astonishing statement that Heaton Colliery had cost £40,000 to win. He estimated that one engine cost £1,200 to build and £400 a year running costs. 'They may have had a bad Fire Engine for 8 or 900£' he explained, 'but I don't know they can have a good one of 32 inches diameter for less than £1,200'.

The introduction of the engine into Scotland was earlier and more extensive than hitherto believed. In the Ayrshire coalfield alone over twenty-five engines had been erected by 1800 and some of these were very early. In particular engines were built at the coastal Stevenston Colliery and their erection and maintenance was made difficult by the lack of local engineering skill. The first engine was ordered from the Proprietors in 1719 and it had an iron cylinder 18in diameter with two iron pit-barrells of 7in diameter. The main beam and pump trees were of elm. The iron boiler was supplied from Frodsham, no doubt by Stonier Parrott. Frodsham was on the River Weaver in Cheshire and before the opening of the Weaver Navigation was the highest navigable port and within reasonable distance of Parrott's work in Staffordshire.

John Meres's difficulty when the order was placed lay in persuading anyone of his people to go to 'wet and slavish Scotland', but Peter Walker who was at Whitehaven was eventually enticed to move to Scotland by November 1719 by a salary of £1 per week. The Scots then offered him £60 a year to manage the collieries but he was persuaded to return to Whitehaven by December 1720 and left the engine unfinished. The work had to be completed by John Potter who then joined Daniel Peck and took over the colliery. The venture failed and a new partnership was formed which immediately built a larger engine in 1721 with a 25in diameter cylinder, which pumped from 20 fathoms. Potter then left the concern, being bought out by Peck who in 1725 built the third engine with a cylinder of 29in diameter in the engine house of the first machine. A further engine was built in 1747 at a cost of around £700.

Another early engine in Scotland was built in 1720 at Elphinstone, East Lothian. Hans Dominik in 1930 published an article in which he quoted the diary of a Richard Reynolds who reported on the work of 'scouring' or machining the bore of the:

> great cylinder of a fire engine for drawing the water from the coal pit at Elphinstone, of a bore twenty eight inches across, and in length nine feet, the same being cast of brass after much discouragement, and the spoiling of three before, which made us much doubt if we could ever succeed in a task of such great magnitude; but being, by reason of the extremity to which the proprietors of the pit were at, having to employ more than fifty horses to discharge the water thereof, much urged to persevere, we give great gratitude to Almighty God, who hath brought us through such fiery tribulations to an efficient termination of our arduous labors.
>
> Having hewed two balks of deal to a suitable shape for the cylinder to lie therein solidly on the earth in the yard, a plumber was procured to cast a lump of lead of about three hundred-weight, which being cast in the cylinder, with a dike of plank and putty either side, did make of it a curve to suit the circumference, by which the scouring was much expedited.
>
> I then fashioned two iron-bars to go around the lead, whereby ropes might be tied, by which the lead might be pulled to and fro by six sturdy and nimble men harnessed to each rope, and by smearing the cylinder with emery and train oil through which the lead was pulled, the circumference of the cylinder on which the lead lay was presently made of a superior smoothness; after which the cylinder being turned a little, and that part made smooth, and so on, until with exquisite pains and much labor the whole circumference was scoured to such a degree of roundness, as to make the longest way across less than the thickness of my little finger greater than the shortest way; which was a matter of much pleasure to

me, as being the best that we so far had any knowledge of; but I was busy casting about in my mind for means as to how it might be in future made better, and I reckoned, for one thing, that I would so fashion the iron bars to which the horses were tied, that they might be laid in the cylinder, and the lead cast on them, and so fasten them firmly.

Details of a licence have survived in the case of an engine built by the Proprietors' agents in the North, John and Abraham Potter, for Andrew Wauchope, owner of the Edmonstone Colliery in Midlothian 'which could not be wrought by reason of water'. This document, dated 1725, is entitled 'A licence granted by the Committee in London appointed and authorised by the Proprietors of the Invention for raising Water by Fire to Andrew Wauchope of Edmonstone Esquire'. The licence granted Wauchope the right to erect one engine with a steam cylinder 9ft long and 28in diameter 'according to the method and manner now used at Elphingstone in Scotland' in return for which right he was to pay an annual royalty of £80 for a term of eight years, that is to say until the patent right expired. This royalty was to be payable quarterly to John Meres at his house in the Apothecaries Hall, Blackfriars, London.

A detailed account dated 1727 shows that the materials for the engine cost Wauchope the very large sum of £1,007 11s 4d, by far the largest single item being the cylinder which, as already mentioned, cost £250. Moreover, unlike Tyneside quotations which are inclusive, this figure does not cover the cost of building the engine house or the labour charges of the engine erectors. The more interesting and revealing items in the account are as follows:

Imprimis. To a cilinder 29 inches diameter with workmanship carried to London and all other charges and expenses	£250	0	0

Other items in this account read:

	£	s	d
To a pestion	9	10	0
To a brass barrel, 7 feet long . .	17	10	0
To Elm pumps at London	53	4	6
To Two cast-mettle barrels 9 foot long and 9 inches diameter and with expenses after them.	41	16	6
To the plumbers bill for lead and lead tap for the boyler with sheet lead and lead pipes	78	10	6
To the timber bought in Yorkshire for the engine with carriage by land and water and freight to Newcastle . .	82	16	0
To two brass buckets and two clacks 9 inches diameter a brass regulator and injection cock and other cocks; sinking vouls [valves], injection caps, snifting vouls and feeding vouls . .	35	5	0
To 44 cwt. 1 qr. 14 lb. of chains, screw work and all other iron work about the engine except the hoops of the pumps at 5d. *per* pound	103	10	0
To plates and revet-iron for making the boyler	75	10	0
To six Swedish plates	6	3	0
To iron-hoops for the pumps with screw bolts and plates for ditto 18 cwt at 4d *per* pound	33	12	0
To Ropes	19	16	0
To Leather	18	13	0
To Cast Metal Bars for the furnace	16	14	0
To Soder	15	10	0
Carriage of the materials from London and Newcastle to Scotland. . . .	22	2	6
John Potter, pains and going and coming upon account of Edmonstone engine	50	0	0

That the engine took twenty weeks to build in 1726 we learn from a separate item covering erectors' wages: 'From July 12th to and with Dec. 1, 1726, being twenty weeks for Ben and Robin at £1 10s. *per* week . . . £30 0s. 0d.'

In addition to all this a separate agreement was entered into with 'John Potter, engineer at Chester in the Street, and Abraham Potter, his brother-german' whereby the engineers were not only to be paid £200 a year to keep the engine going but were to receive half the clear profits of the colliery during the eight years' term of the first agreement with the Proprietors. Abraham Potter was also appointed steward and factor of the colliery. Should the brothers Potter fail to keep the workings clear of water they were entitled to take away all the materials they had furnished and to be paid a reasonable allowance for their pains and charges. It is surprising that any colliery owner would submit to such onerous terms but, having a drowned-out pit, the unfortunate Andrew Wauchope was faced with no alternative other than the abandonment of his enterprise.

The royalties demanded by the Proprietors were as irksome to engine owners as were those levied by Boulton and Watt at a later period. Griff Colliery, which had two engines at work in 1725, closed down a few months after the death of Sir Richard Newdigate in July 1727 and in 1731 the engines were dismantled and sold. The main reason for closure was that the more accessible seams had been worked out, making it impossible for Griff to compete with the neighbouring pit at Hawkesbury, but the burden of royalty payments to the Proprietors was an important factor in the

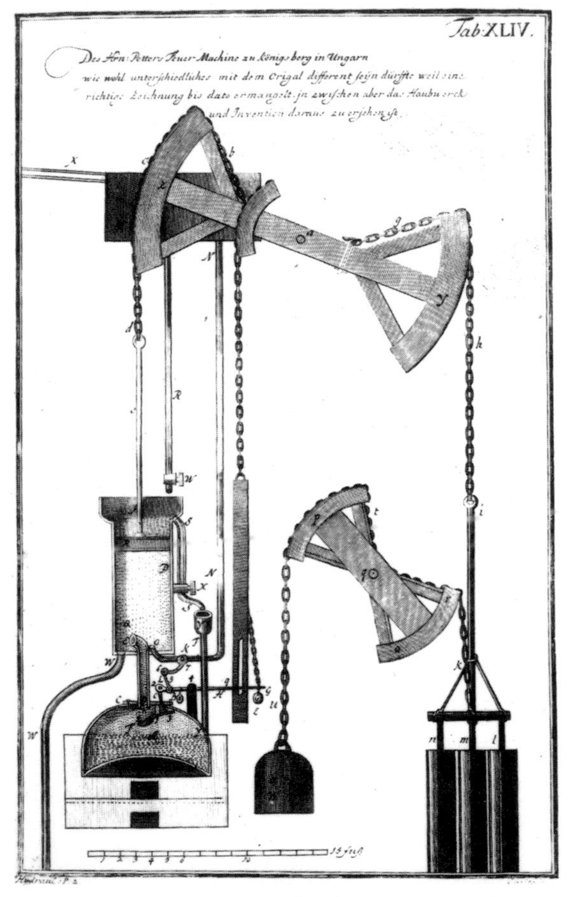

27 Schematic drawing of the 1722 engine at Königsberg, Hungary by Jacob Leupold, *Theatrum Machinarium Hydraulicarium* Vol I 1725. Leupold had not seen the engine and shows a separate balance beam attached to the pump rods.

decision and in 1728 a re-opening was envisaged when the patent expired. A document of this period in the Newdigate papers argues that: 'The fire engine Act will then be expired and Griff engines which now pays 300 li. p. an. rent will then be rent free. In which article there will be saved to the family £1,500 0s. 0d.'

From such records as these it is easy to understand why, despite its overwhelming advantages, the introduction of the Newcomen engine, although amazingly rapid, might have been even faster but for the fees which had to be paid to the Proprietors of the Patent. Just as in the last decades of the eighteenth century mine owners clung stubbornly to their Newcomen or 'common' engines, as they were then called, rather than pay tribute to Boulton and Watt, so in the first decades there must have been many who continued to struggle along with water or horse-powered pumps and looked forward to the day when the Savery patent monopoly would expire. To judge by two verse broadsides which appeared in 1720-1, the time of the South Sea Bubble, the fortunes to be made by the engine from mining were sometimes doubtful. *The Broken Stock-Jobbers: Epilogue by a Loser* (1720), had this to say:

> Why must my stupid Fancy e'er admire
> The way of raising Water up by Fire?
> That cursed Engine pump'd my Pockets dry,
> And left no Fire to warm my fingers by.

The Bubbler's Mirrour, or England's Folly (1721), expressed an equally dim view of the financial prospects of the engine business:

> Come all ye Culls my Water Engine buy
> To Pump your flooded Mines, and Coal-pits dry.
> Some projects are all Wind, but ours is Water,
> And tho' at present low may rise here a'ter.

Yet the sheer merit of Newcomen's invention was such that the engine continued not only to make rapid progress at home but began to sell abroad.

There were a number of mining areas on the Continent where drainage problems were just as pressing and expensive as in Britain, particularly in Belgium and Slovakia. The mine owners there were naturally interested in the Newcomen engine. A number of Continentals visited Britain and at least one disguised himself as an artisan to discover the secrets of the engine. By 1719 mining

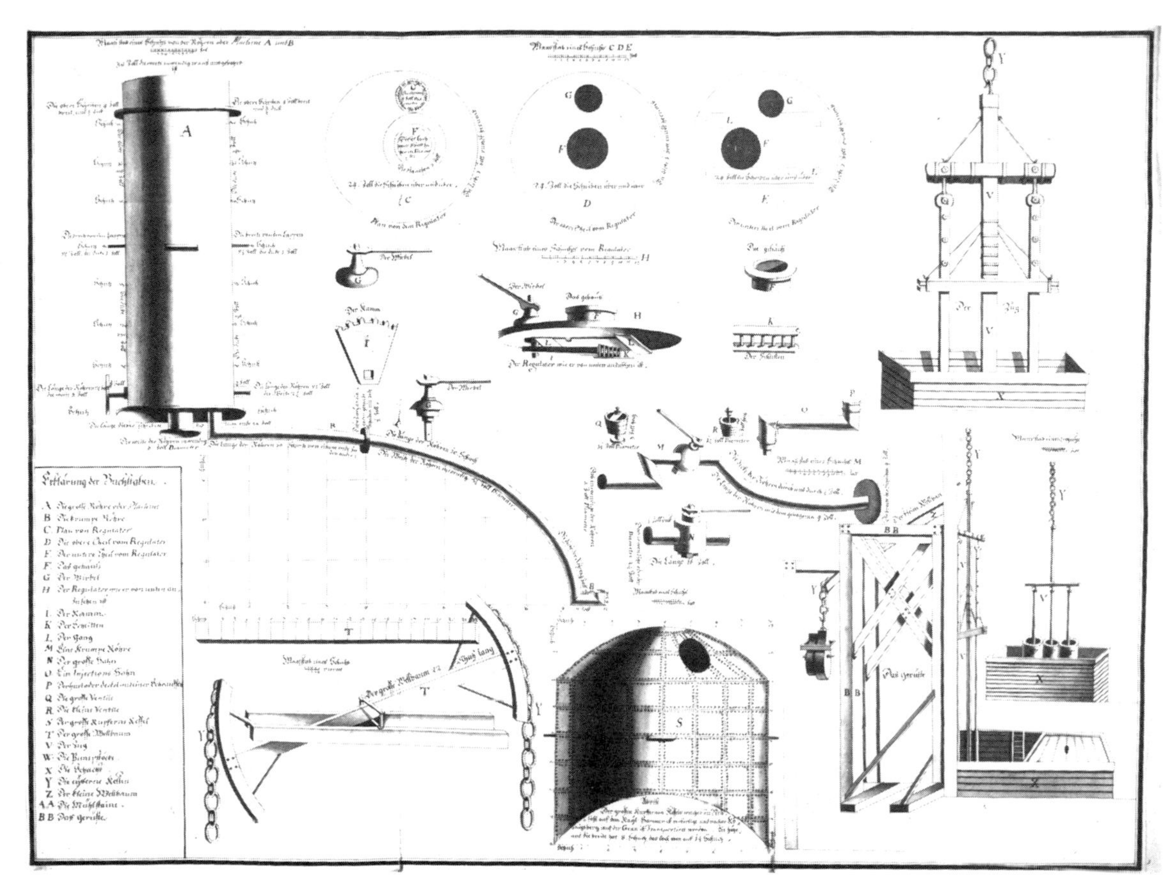

28 Detailed drawing presented by A. D. von Schönström (1692-1745) in 1725 to Dr Daniel von Guldenberg, Hanoverian envoy in Vienna. This is the earliest drawing of a continental engine and also indicates the second balance beam connected directly to the pump rods. It shows also a plain boiler without flanges similar to the Dudley Castle engine.

interests in both Vienna and Liège were negotiating with English mechanics. The first to arrive on the Continent were Isaac Potter who was in Vienna by the summer of 1720 and John O'Kelly who was in Liège by September of the same year.

O'Kelly was the first to complete an engine, for which he had signed a contract on 26 March 1720 with three entrepreneurs of Liège: Wanzoulle, Van Stein and d'Aubée, O'Kelly himself acting as engineer. The engine was built at Jemeppe-sur-Meuse near Liège with a 25in diameter cylinder supplied from England and was completed by February 1721. The pit was 327ft deep and the engine lifted the water 65ft to an adit using 9in diameter pipes. There were difficulties over finance and O'Kelly sold out in 1723. The engine was moved at least once and may have been re-erected at Péry near Groumet where an engine was reported at work in 1725. Reference has already been made to O'Kelly who returned to London in 1725.

Isaac Potter and his party, consisting of his servant Lumley, Lumley's daughter and Potter's assistant engineer Pierre Sabathery, arrived in Vienna in June 1720. After negotiation Potter agreed to drain a long drowned-out mine at Königsberg (Nová Bana). A bilingual document was drawn up in German and English on 1 September 1721 in which Isaac Potter gave full power to two of his associates, de la Haye and Costebadie to execute 'all his affairs in general, but in particular everything concerning the Construction and Erecting of the Fire-Machine'. The names of Potter's associates are French or Walloon so he may have recruited them en route.

29 General view of the Königsberg engine also presented by von Schönström in 1725. This drawing also indicates the single main beam.

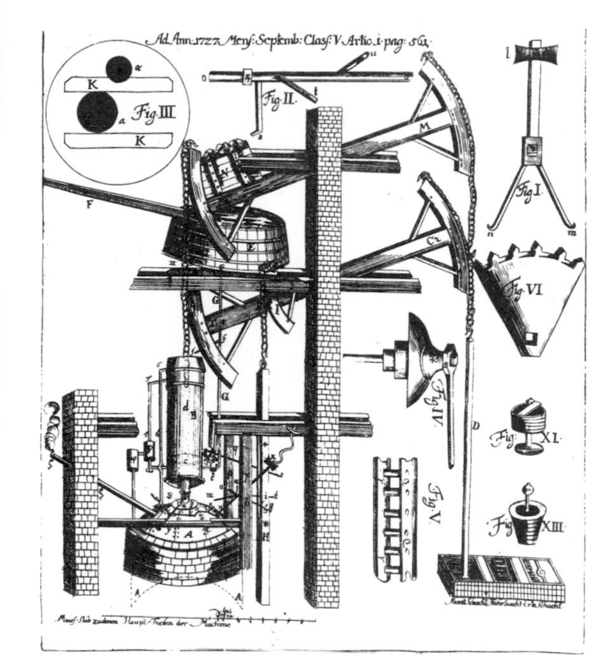

30 The engine at Königsberg as drawn by G. M. Kortum *Miscellanea-physico-medico-mathematica* in 1727. This drawing shows the use of a double beam and the inference is that the engine was modified from its original form.

The engine had a brass cylinder 30in diameter and 8ft long and drew 270 gallons per minute from a depth of 460ft.

The engine sited at Königsberg was sixteen English miles from Schemnitz (Chemnitz) in the mining district of what is now Czechoslovakia. The first trials took place in 1722 and following improvements made by J. E. Fischer Von Erlach (1692-1742) the engine was stated to have worked properly in February 1724. Von Erlach had become familiar with the Newcomen engine during a tour of England between 1718 and 1720. The engine worked until 1729 and stood idle until 1733. It was dismantled during 1734 and re-erected at Hodritz a little to the west of Windschact. Jacob Leupold in his *Theatrum Machinarium Hydraulicarum* (1725) gives a description and drawing of the engine which indicates a balance bob for the pump rods; this is probably the first use of such a system.

A Swedish nobleman A.D. von Schönström (1692-1745) made a study tour of mining districts in Europe starting in 1724. Coming to Vienna in 1725 directly from Königsberg, he handed two drawings to Dr Daniel von Guldenberg, the Hanoverian envoy in Vienna. These drawings indicate that the engine had one main balance

31 Details of the Königsberg engine by Kortum in 1727. The steam valve is worked by a toothed sector and is clearly the same engine as that in the von Schönström print.

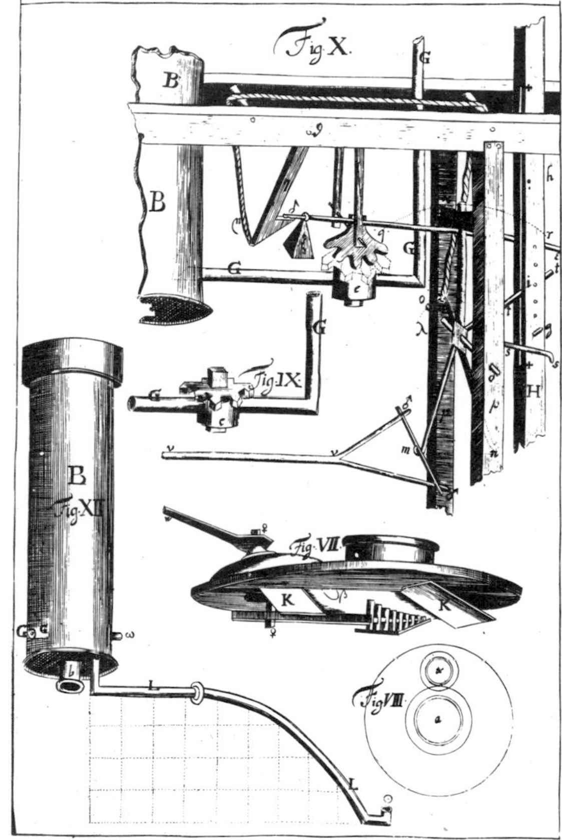

beam and a separate one connected directly to the pump rods, supporting Leupold's drawing.

A meticulous description and a general view of the engine with a number of details were published in 1727 by G. M. Kortum in *Miscellanea Physico-Medico-Mathematica* from which the method of counterbalancing the weight of the pump rods by a second balance beam erected over the main engine beam can be seen. The injection valve was operated by a toothed sector. These changes may well have been the Von Erlach improvements.

An engine at least as early as the Königsberg machine appears to have been built at Cassel in Germany. Evidence for this engine is much less positive. Some writers have doubted whether it was built, claiming confusion with a Savery engine built there in 1715, but there is a report of an interview in 1727 with Fischer Von Erlach in which he claims to have built the engine. It is most unlikely that a man of his standing would make such claim if it were untrue. Furthermore, J. N. S. Allamand in *Oeuvres Philosophiques et Mathematiques de Mr. G. J. 'sGravesande* (Amsterdam 1774), refers to a visit he paid to Cassel in 1721 to advise on the installation of new machines.

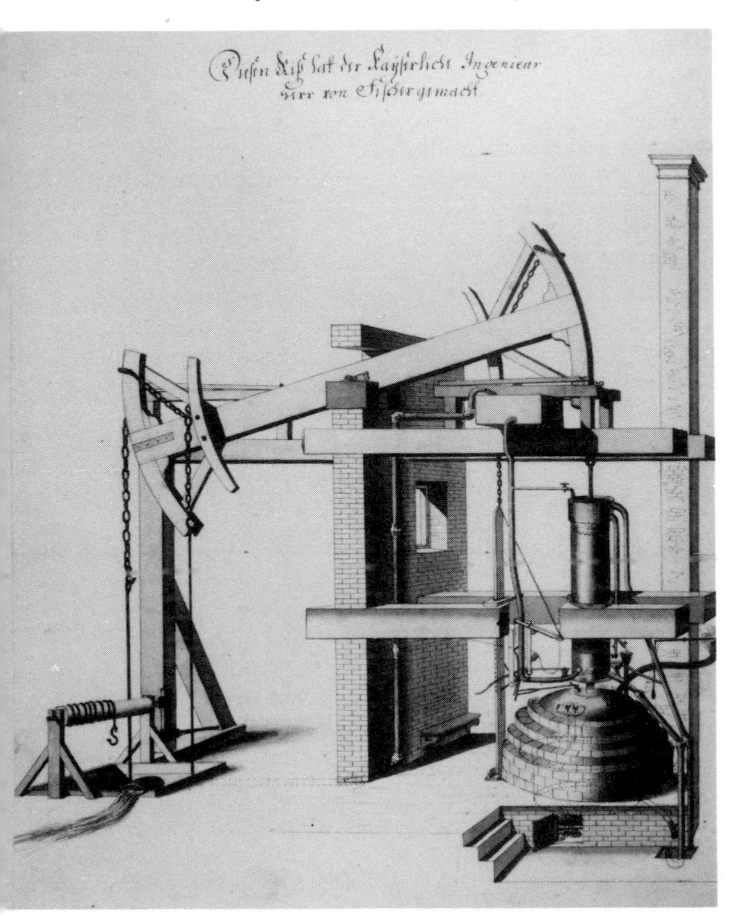

32 Drawing of the Könisberg engine stated to be by J. E. F. von Erlach. However this is quite different from the other drawings and seems to be based on an English original.

Von Erlach certainly built an engine in Vienna about 1722/3 in the gardens of Prince F. A. Von Schwarzenburg following the signing of a contract in early 1722. This engine had a cylinder 24in diameter 9ft long with a weight of 1,200lb. The sloping rising-main to feed the header tank of the fountains reached a height of 75ft above the sump from which the engine pumped and working at 15 to 16 strokes per minute it lifted 8,000 gallons per hour. The piston was sealed with leather. Von Erlach was responsible for an improved boiler design here and led the flue twice round the boiler before it reached the chimney. After an overhaul in 1747 the engine continued to operate until 1770.

Triewald referred in his book to a letter from the Viennese gunfounder Leopold in which he stated that a decision had been taken to replace 500 horses in the Schemnitz district of Hungary with five fire engines, following the success of the Königsberg engine. An agreement was signed with Fischer von Erlach on 8 June 1732 for the building of two engines at Windshacht and for a further two in June of 1734. The first two were completed by 1733 and the second by 1735. Isaac Potter acted as engineer during this work and was to die in Hungary in 1735.

All four engines had cylinders 32½ inch diameter by 9ft long and beams 24ft long. Operating at eight strokes per minute they lifted 115 gallons per minute. Two of the engines lifted from a depth of 520ft and the other two from the great depth of 900ft.

It would seem that these engines were fitted with two balance beams reflecting the depth from which they pumped and the weight of the pump rods which had to be counter-balanced. They were also unique in that the cylinder was offset from the centre line of the boiler.

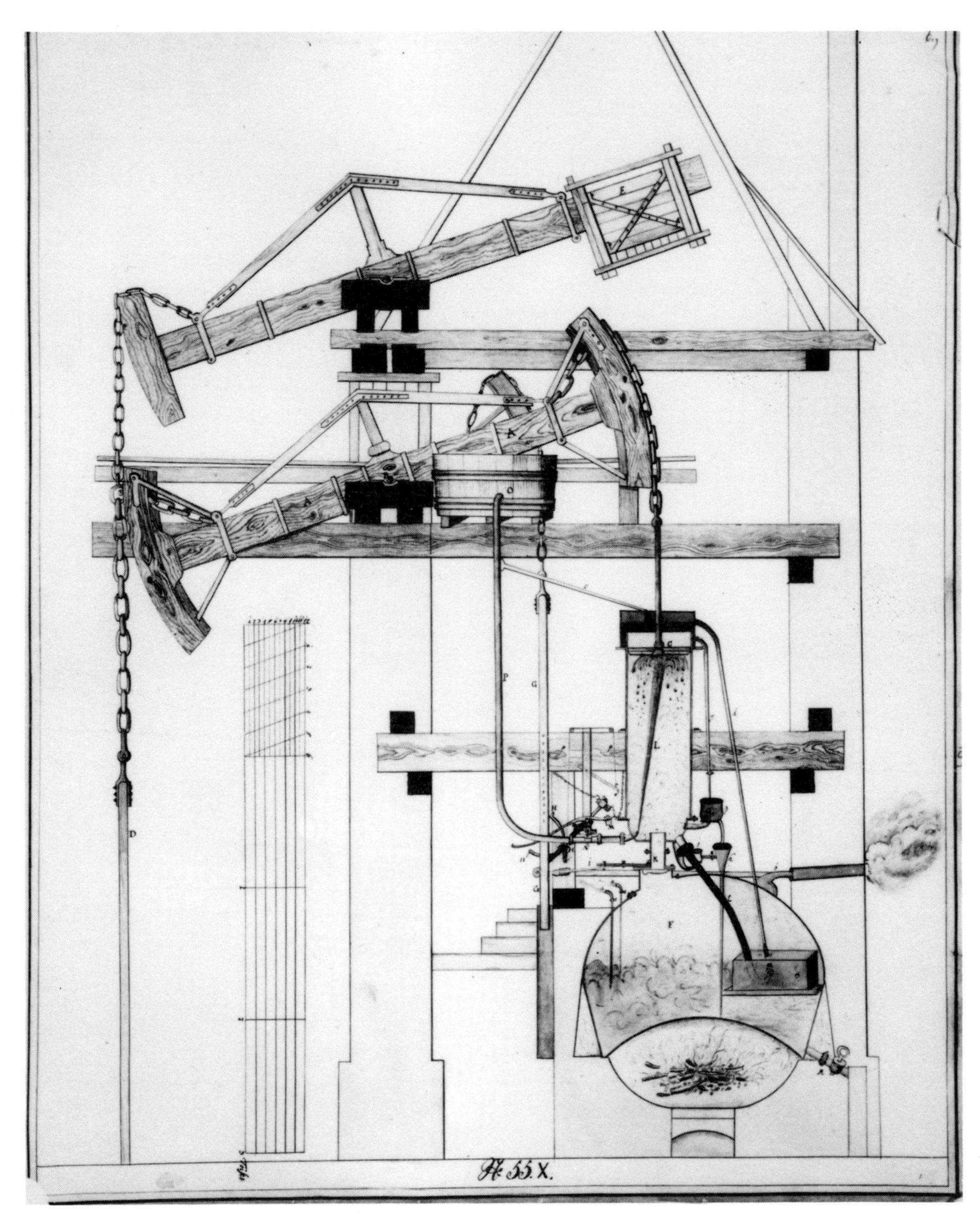

33 The double beam engines of 1732-5 at Schemnitz, Hungary as drawn by Andreas Fritsch. Isaac Potter was the engineer during the building of the engines and the upper beam was to counter-balance the great weight of the pump rods when pumping from 520 and 900 feet. The engines were unique in that the cylinder was off-set from the centre line of the boiler.

In 1737 the last two engines at Schemnitz were removed and set up to pump the same shaft as the first two. Their place was taken by a larger engine of 36in diameter raising water from the fantastic depth of 1,255ft to an adit 212ft above.

It was not until 1758 that the next engine was erected in the area and this was built by Joseph Höll who had perhaps been trained by Potter. This drained the Konigsegger Shaft through flat rods and had a 36in cylinder.

Drawings of these engines exist and were made by an Hungarian student, Andreas Fritsch. They were found in the library of Batthyany in Keszthely, Hungary and are at present in the National Archives in Budapest. Other drawings exist, for it seems the Prussian King — Frederick the Great — wished to buy engines for the mines in his country. Schöner, a mining engineer at Schemnitz, had made many fine drawings for official use and another technician, Aschenbrier copied these drawings for the King. The drawings were ultimately in the library of Charlottenburg, Germany and were destroyed by fire during World War II, but microfilm copies still exist.

A drawing of an engine at Schemnitz is shown in considerable detail by Ch. Delius, *Anleitung zu der Bergbaukunst* (Wein 1773), but here the cylinder is centred above the boiler.

Reference has already been made to documents in the Swedish Royal Archives containing particulars of the offer made by John O'Kelly to construct an engine for the Swedish Department of Mines. However O'Kelly's Swedish project failed to materialise, and the way was clear for Triewald to build the first engine in Sweden in 1727 at the Dannemora Mine.

Triewald at first tried to obtain a 36in diameter iron cylinder from England, but meeting no success he approached Gerhard Meyer, Jun, in Sweden and in December 1727 gave him an order for the cylinder and the piston. Meyer was a member of a family who for 150 years had managed the Royal Gun Foundry in Stockholm and himself was appointed Royal Gun Founder at the age of 24 in 1728. Meyer had visited England in 1724 where he studied gun and cylinder casting as well as general iron casting. Later he visited Leopold at the gun foundry in Vienna and it is possible that he may have seen the Königsberg engine in Hungary.

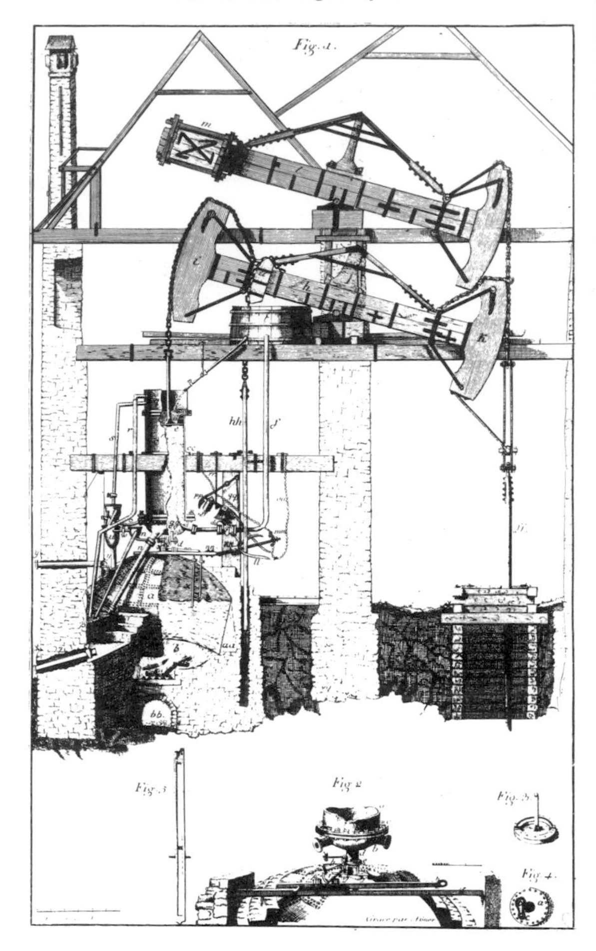

34 A further engine at Schemnitz, Hungary, from C. T. Delius, *Anleitung zu der Bergbaukunst* 1773. The engine also has a double beam but the position of the cylinder is conventional. The whole engine including the pump rods and shaft top is within an engine house.

The cylinder, an alloy of 100 parts copper and 10 parts tin, then called gun-metal, had cast upon it the names of Marten Triewald and Gerhard Meyer with the date 1728 and these were surrounded by a decorative embellishment. Triewald stated that the cylinder was the most perfect he had seen and the piston was an excellent fit. This was in fact the first masterpiece of young Meyer who later became renowned for the casting of many fine statues, monuments and bells. The engine house, which still stands at Dannemora,

35 Marten Triewald. An engraving by E. Geringius after a painting by Schröder, published in 1735.

36 The engine at Dannemora, Sweden built by Marten Triewald in 1727/8. The cylinder of 36in diameter was supplied by Gerhard Meyer of Sweden.

has granite block walls and a tiled roof. It was built by the mine workmen who also constructed on the site the heavy balance of six pine logs with four oak quadrants. The boiler was built in Stockholm by a coppersmith and installed in the engine house before the cylinder was erected. A Bill of Lading confirms that some cast iron items were supplied from England.

The whole of the construction was made in accordance with detailed designs and drawings which are amongst the earliest engineering drawings extant for a steam engine. They are preserved in the archives of the 700-year-old Stora Kopparberg Co at Falun.

Unfortunately the engine did not prove the success Triewald hoped and was said to be 'very unreliable in its way of working. Something seems always to go to pieces and has to be immediately repaired.' Triewald had also designed machinery to operate from the engine to raise the ore, but this was not completed until 1730 and was also not a success. In 1735 the pump house had collapsed

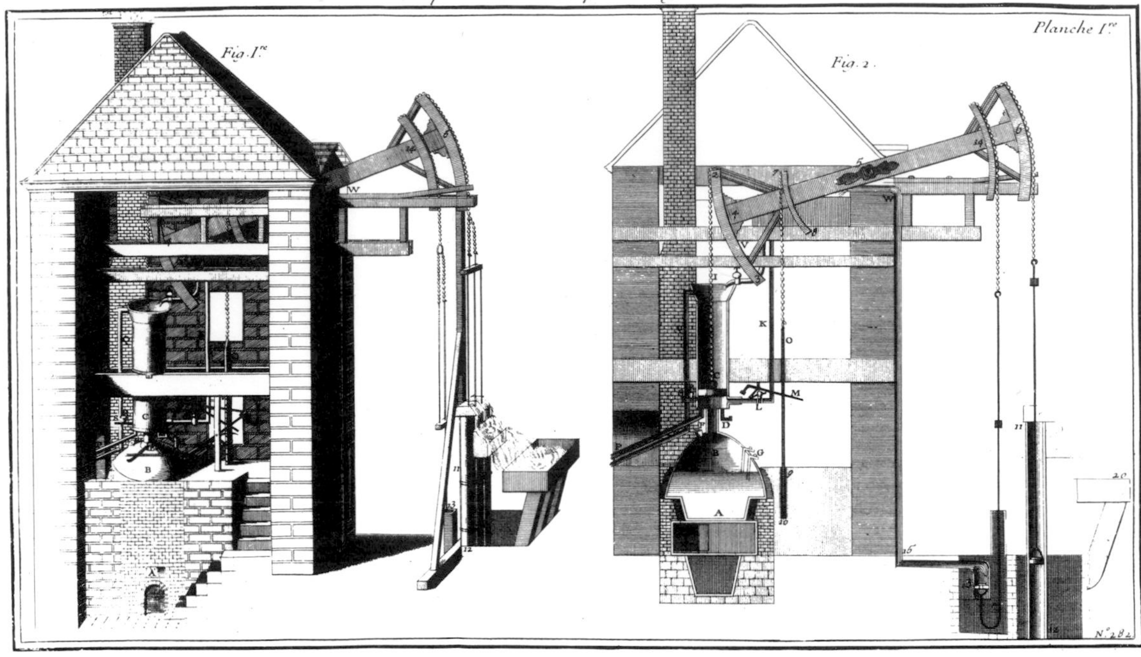

37 The engine built at Passy, near Paris in 1726 by 'May and Meyer'. From *Machines et Inventions approuvées par l'academie Royales des Sciences,* Tom IV, 1735. The builder was John May, accompanied by John Meres, who died in France.

and lightning struck the engine house in 1736 destroying part of the roof and splitting the chimney. The use of the engine then seems to have been abandoned, but the parts were not sold until 1773.

It appears that an engine was to be built in Spain in the 1720s to pump water from the Tagus to the city of Toledo. This was a venture of an English company which had no connection with O'Kelly and obtained a concession in 1722. The engine is said to have been built by Richard Jones Esq, of London but whether the project was ever completed is uncertain.

The first Newcomen engine in France was built at Passy, near Paris in 1726 for the purpose of pumping water from the Seine to the city. The engine caused a considerable stir in Paris and was very fully described and illustrated in *Machines et Inventions approuvées par L'Academie Royal des Sciences*, IV, 1726 (published in 1735) and is said to have been capable of raising 20,925 Muids, (8,286,300 gallons) of water in 24 hours, a height of 30ft. The publication credits the engine to 'Mm Mey et Meyer' but the two French engineers d'Onzembrais (Ons-en-Bray) and de Réaumur

38 Detail of the valve gear on the Passy engine.

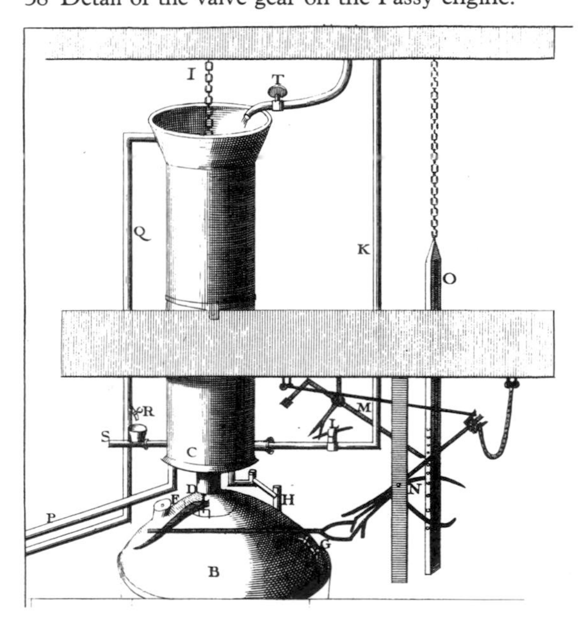

who were named by the academy to examine and report on it refer only to the English M. Jean May as the builder. On 11 May 1726 in describing the engine they state that it is 'extremely ingenious in all particulars and one which does great honour to the genius of its Inventors'.

John May pressed the French authorities for a special priviledge for the erection of engines in France. At the same time the authorities had to consider claims from a French architect Germain Bosfrand (1667-1754) who offered both a Savery and a Newcomen type engine. The Newcomen machine had been built by 'the brother of the engineer Potter of Durham who had built the Königsberg engine', and had been installed at Cachan near to d'Areueil by Bosfrand. They also considered claims by another Englishman, Baker, who proposed to build engines on the Newcomen principle. John May won the day with the support of Réaumur and was granted a priviledge for 20 years from 3 July 1727. The M. Meyer is identified as John Meres who was in France at the time and in fact died in Paris in May/June 1726 shortly after the Passy engine was completed.

George Saunders who had been an assistant to O'Kelly at Liège arrived in Vedrin, near Namur, Belgium in 1730 and almost certainly built four engines to drain the silver-lead mine there. They were completed in 1730, '35, '38 and '40.

The next engine in France was built in 1732 at Fresnes near Condé on the Nord coalfield and was also completed by Saunders who improved on the

39 Engine designed by M. De Bosfrand in France 1727. His valve operation was by toothed sector. From *Machines et Inventions approuvées par l'academie Royal des Sciences*, Tom IV, 1735.

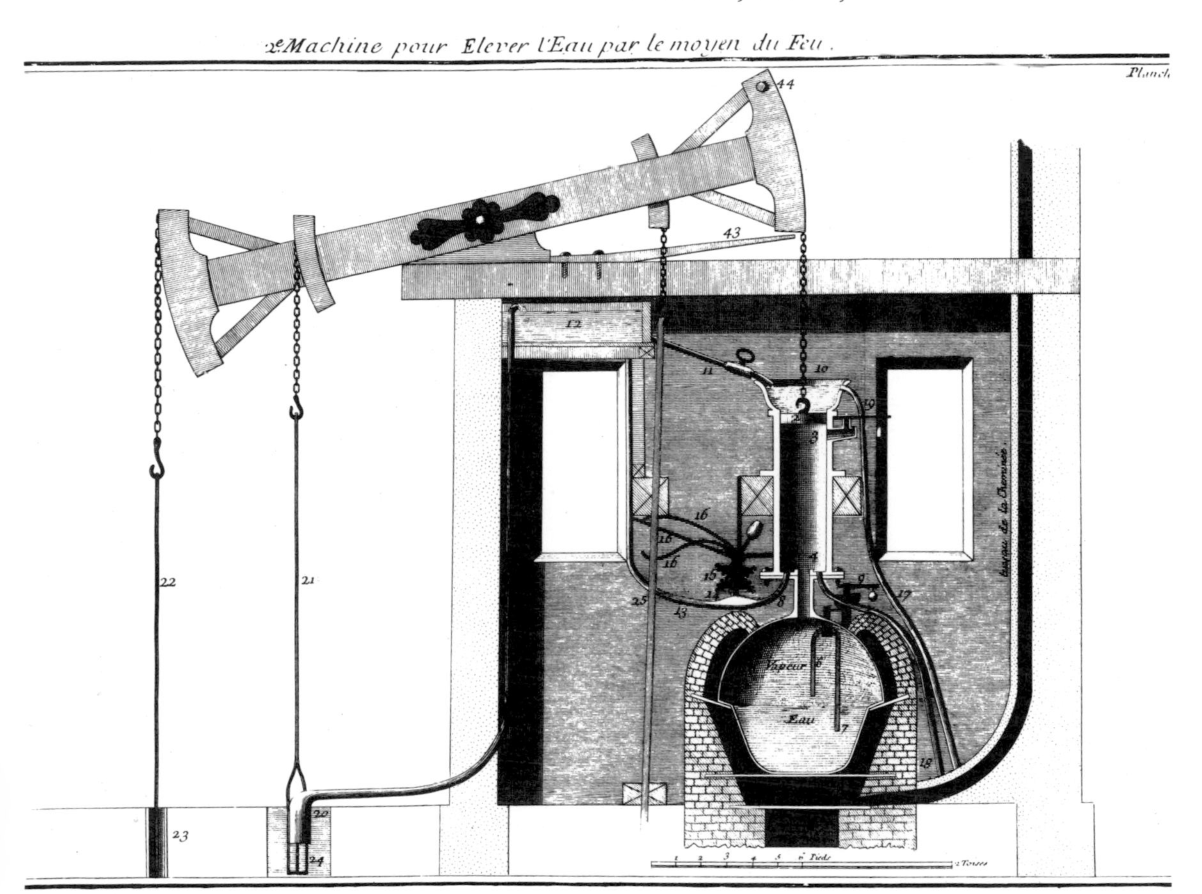

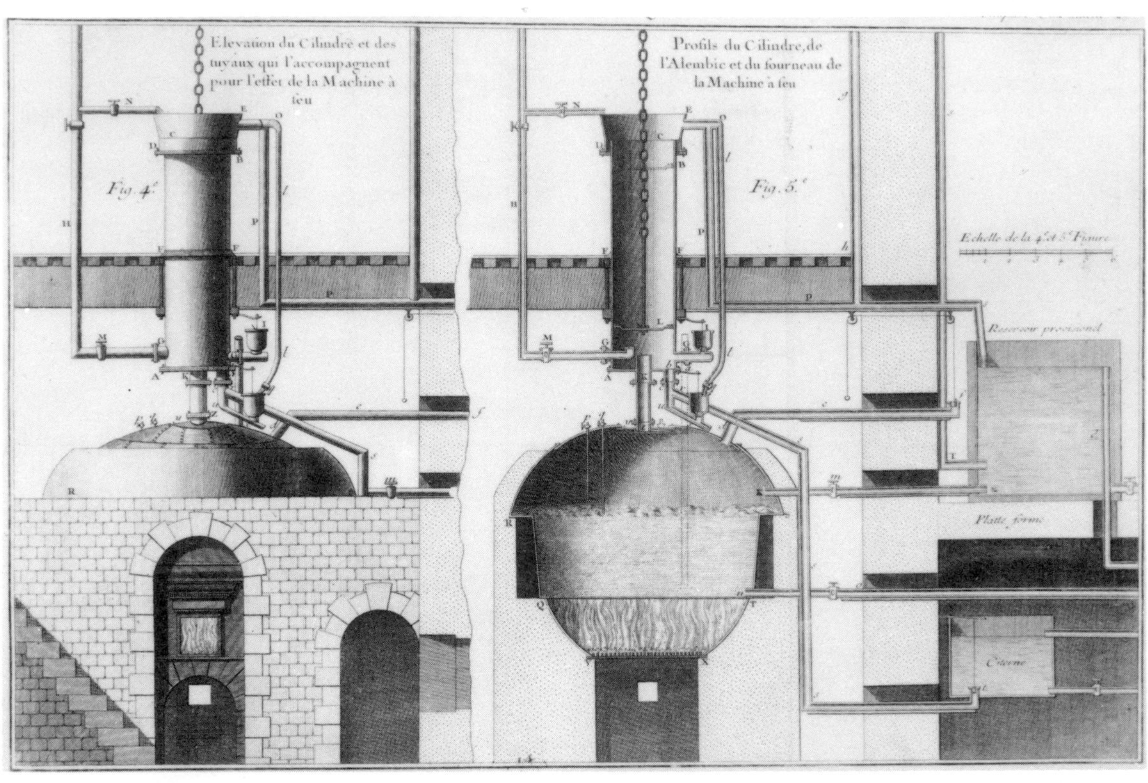
Elevation du Cilindre et des tuyaux qui l'accompagnent pour l'effet de la Machine à feu
Profils du Cilindre, de l'Alembic et du fourneau de la Machine à feu
Fig. 4.e
Fig. 5.e
Echelle de la 4.e et 5.e Figure
Reservoir provisionel
Platte forme
Citerne

Dessein général des principales parties d'une Machine qui eleve l'eau, par l'action du Feu.
Fig. 1.re
Fig. 2.e
Fig. 3.e
Reservoir provisionel
Chapiteau de l'Alembic
Echelle de la premiere et troisieme figure
Echelle de la seconde figure

boiler design. These mines by 1750 had five engines at work. Saunders also built an engine at Lodelinsart (Charleroi) in 1735 and he trained a number of Walloon assistants who after his death about 1745 erected a great number of engines for the coal owners of Borinage (Imperial Hainaut – West Mons Region), Charleroi and Liège. During the period 1740-90 it is stated that over one hundred engines were built. An innovation in the French Hainaut region was the use of portable engines fixed in wooden frames for use during shaft sinking. Another engine was built at Littry about 1744-6 which was apparently not a success. There are still some parts of a later type of engine remaining in an engine house at Littry.

As a water-supply pump Newcomen's engine had invaded the capitals of England and France simultaneously for it was in the year 1726, that an engine was completed at the York Buildings waterworks which had been established on the bank of the Thames in 1676 on a site at the lower end of Villiers Street, Strand. It will be recalled that Savery installed one of his engines at York Buildings. John Farey in his *Treatise on the Steam Engine* refers to an engraving in *Voyages de M. de la Motraye* (1732), Vol III, which shows the Savery and Newcomen engines at York Buildings standing side by side and pumping into a common cistern. Farey infers from this that the two engines did work together for a time, but this seems extremely unlikely in view of the difficulty which Savery experienced in getting his machine to withstand the steam pressure necessary to give the required lift.

Like its opposite number in Paris the York Buildings Newcomen engine attracted much notice, though in this case not always favourable. An alleged representation of the engine, engraved by Sutton Nicholls in 1725, was produced for sale, perhaps at the instigation of the Proprietors and its publication was announced in the following advertisement which appeared in *The Daily Post* for 28 March 1726:

> Whereas the famous and useful engine for raising water by fire will soon be at work at the York Buildings Waterworks, in the Strand; there will on Wednesday next be published a true draft of the same, as it is at work in the several parts of great Britain, and proper explanation. Sold by John King at the Globe in the Poultry. Price One Shilling.

F. D. Klingender in *Art & the Industrial Revolution* 1947 tells us that the publication of the engraving was a counterblast to a lampoon which was first published in December 1725, and reprinted in 1726. Klingender quotes the title:

> The York Buildings Dragon; or, a full and true account of a most horrid and barbarous murder intended to be committed on Monday the 14th day of Febr. next (being Valentines-day) on the Bodies, goods and name of the greatest Part of His Majesty's

42 A Science Museum model of the Fresnes engine.

40 Elevation and section of the 1732 engine at Fresnes, near Condé on the Nord Coalfield in France, illustrated by B. F. de Belidor in *Architecture Hydraulique*, Tom 11, 1739.

41 General arrangement and details of the valve gear and boiler of the Fresnes engine by Belidor.

43 The engine engraved by Sutton Nicholls in 1725, copies of which were on sale to the public. It is said to represent the engine at York Buildings. The 35½in diameter brass cylinder was sold in 1732 to Sir James Lowther at Whitehaven, and a second cylinder to the London Lead Company.

> Liege subjects, dwelling and inhabiting between Temple-Bar in the East, and St. James in the West, and between Hungerford Market in the South and St. Marylabonne in the North by a sett of evil minded Persons, who (by the Instigations of Plutus, and not having the fear of several Lords, Knights and Gentlemen before their eyes) do assemble twice a week, to carry on their wicked purposes, in a private room over a stable by the Thames side, in a remote corner of the Town. The second edition, augmented by almost Half London.

Whether the 'sett of evil minded persons' referred to the Proprietors of the Patent or to the proprietors of the York Buildings Company or to both is not clear. When fed with live coals by 'a Lancashire wizzard, with long black hair and grim visage . . . and a Welchman bred on top of Penmaenmaur', says the author, 'the monster will

44 A drawing showing the two engines at York buildings from *The Voyages and Travels of M. De la Motraye*, Vol III, 1732. The similarity of the Newcomen Engine to the Sutton Nicholls engraving will be noted. The Savery engine is drawn to a different scale and it is unlikely the two engines were worked at the same time.

45 A drawing of a Newcomen waterworks engine. From S. Switzer, *Hydrostaticks and Hydraulicks* 1729.

A DRAUGHT in PERSPECTIVE of an ENGINE for RAISEING WATER by FIRE.
Fig. 2.
Fig. 3.
Toms sculp.

clap his wings several times successively with prodigious force, and so terrible will be the noise thereof, that it will be heard as far as Calais, if the wind set right'. He will then suck from the Thames 'such a prodigious quantity of water that barges will never be able to go through the bridges'; then, being 'of a *huffing*, *snuffing* temper, he will dart out of his nostrils perpendicularly up to the skies two such vast, dense, and opake columns of smoke, that those who live in the Borough will hardly see the sun at noonday'. It was certain that the monster would finally poison the populace with the venom which he would draw from the river 'through a long proboscis, something like an elephant's trunk'.

There is a remarkable resemblance between this extravaganza and the pamphlet prophecies of death and disaster which accompanied the birth of the steam locomotive more than a century later. The latter were inspired by the coaching industry which the new railways threatened and it has been suggested that the York Buildings Company's great rival, the New River Company, may have been behind this earlier effort. The historical parallel may be taken further. Like the locomotive, Newcomen's engine remained safe from attack and was virtually ignored so long as its use was confined to the industrial areas of the Midlands and the North, which were beyond the pale of polite society. But as soon as these inventions were applied to public service in the South and dared to invade London they became targets for attack.

The York Buildings engine seems to have worked very successfully, but that extravagant appetite for fuel, which mattered little to the owner of a coal mine threatened with inundation, told heavily where coal had to be brought by sea from Newcastle. Dr Allen, a friend of Newcomen, stated that the engine consumed £1,000 worth of coal a year and on top of this there were the inevitable royalties to be paid to the Proprietors. The waterworks company soon got into serious financial difficulties and in 1731 the use of the engine was discontinued. At the end of the following year they still owed the Proprietors £787 10s. A 35½in diameter brass cylinder was sold in 1732 to Sir James Lowther at Whitehaven where it was first fitted to the Saltom engine, and another 35½in diameter cylinder and piston was sold to the London Lead Company for their engine at Trelogan. Whether this second cylinder was a spare or from a second engine is as yet uncertain. It was at least a decade before the York Buildings engine was superseded by a larger machine of Newcomen type and in the meantime the company presumably fell back upon the horse-driven pumps, such as they had used before their experiments with steam engines began. A paragraph in *All Alive and Merry; or the London Daily Post* for 18 April 1741 described the slumbering monster thus:

> There is a famous machine in York Buildings, which was erected to force water by means of fire, thro' pipes laid for that purpose into several parts of the town, and it was carry'd on for some time to effect; but the charge of working it, and some other reasons concurring, made its proprietors, the York Buildings Company lay aside the design; and no doubt but the inhabitants in its neighbourhood are very glad of it; for its working, which was by sea-coal, was attended with so much smoke, that it not only must pollute the air thereabouts, but spoil the furniture.

How unpolluted an atmosphere must London's inhabitants have enjoyed when the smoke from one boiler chimney could provoke so much adverse comment. Their descendants would become inured, first to a perpetual pall of coal smoke and then to the fumes from thousands of motor vehicles exhausts.

The account books of the Coalbrookdale Company, which record the supply of iron cylinders and other engine parts, show that the industrial Midlands, where the Newcomen engine was born, must have continued to be the Proprietors' best market. This fruitful area was soon extended northwards to embrace the lead-mining district of Derbyshire. The first engine in Derbyshire had been built at George Sparrow's Yatestoop mine near Winster in 1716. The second engine there was built in 1719 and by 1724 another engine was at work. They are described with remarkable lucidity in the diary of the Reverend James Clegg, a nonconformist minister of Chapel-en-le-Frith who visited the district in the autumn of that year.

> September 28th 1730 — I set out with Mr Tricket to see some remarkables in several parts of ye Peak. Called at Money Ash; went by Middleton, nr

Youlgreave; came to Winster about noon. Saw 3 curious Engines at work there, which by ye force of fire heating water to vapour a prodigious weight of water was raised from a very great depth, and a vast quantity of lead ore laid dry. The hott vapour ascends from an iron pan, close covered, through a brass cylinder fixed to the top, and by its expanding force [sic] raised one end of the engine, which is brought down again by the sudden introduction of a dash of cold water into ye same cylinder which condenseth the vapour. Thus the hott vapour and cold water act by turns, and give ye clearest demonstration of ye mighty elastic force of air . . .

Meanwhile, despite the handicap of its high fuel consumption particularly in districts far from coal supplies, the engine continued to make headway in the Cornish mines through the agency of Joseph Hornblower. It will be recalled that the first engine to be built by Hornblower in Cornwall was at Wheal Rose near Truro in 1725 and he went on to build a second at Wheal Busy and a third at Polgooth. It is said that by 1727 there were five engines in Cornwall, though it is not clear whether this figure refers to engines built by Hornblower or whether it includes the earlier engines at Wheal Vor and Wheal Fortune.

There was a tradition in the Hornblower family that Joseph worked in Cornwall under the direct supervision of Thomas Newcomen. We know from Newcomen's statement in a Chancery case that he was away from home for the whole of 1718 in Cornwall and it would seem most likely that he would be involved with engine building and probably this would be the Wheal Fortune engine. Its exact date of building is not confirmed and it could be that Hornblower assisted in its construction. The engine is certainly the last that Newcomen is known to have been involved with directly.

We would dearly like to know more of what Newcomen was doing during the last years of his life. Although a note in the Dartmouth Baptist register states that he was acting as their pastor, the contemporary evidence shows that he continued to be concerned with engineering and to have spent much time in London. He wrote from Dartmouth in 1725 to Chief Justice King in connection with a wind engine and mentioned his nephews Elias and George. He refers to the death of their agent Mr Calley and this was most likely to be the elder son of his original partner.

The letter which Newcomen wrote from London to his wife in December 1727 shows that he was staying with Mr and Mrs Edward Wallin and from internal evidence indicates that he had been away from home for some while. His son Elias had also been in London but had returned to Devon.

The Chancery case concerning his financial affairs was commenced in 1728 and then as always when describing himself, he wrote, 'Thomas Newcomen, Ironmonger of Dartmouth'. The last contemporary reference before his death concerns London, where he was at the foundry of Harrison and Waylett on 29 January 1728/9.

Newcomen did not live to see his engine freed from the control of the Proprietors by the lapse of Savery's patent. He died at Edward Wallin's house in London — of a fever, according to tradition — on 5 August 1729 when in his sixty-seventh year. A nephew at once despatched the news of his death to the great inventor's eldest son, Thomas, the serge-maker of Taunton. This letter reads:

London, 5th August, 1729

Dear Coz,

I am sorry that I should be the messenger of the ill tho' expected news of my Uncle's, yr Father's Death, for this morning about 6 of the clock it pleas'd the Almighty to take him out of this miserable world, doubtless to enjoy a far better. Indeed Mr. Wallin very prudently ordered the greatest care to be taken of him that possibly could. He had the advice of two Skilful Physitians every day. He had a careful Nurse continually with him, and one or two sat up with him every night. He was very submissive and patient all his Illness, and departed without a sigh or a groan, as if He had been fallen asleep. If you have any business here, which I can be any possible means do for you, I desire you would send word of it to

Sr Yr Sincere Friend & Servt,

JOHN NEWCOMEN.

Please to direct me at
Mr. Thos. Dugdale's, Attorney at law,
in Token House Yard, London.
To Mr. Thomas Newcomen,
in Taunton,
Somerset.

The Monthly Chronicle (vol II, p169), carried a brief obituary note which reads: 'About the same time (7 August 1729) died Mr. Thomas Newcomen, sole inventor of that surprising

My Lord

It having pleased God in his Allwise Providence in some measure to dissappoint us by the Death of Mr John Calley, who was sent to Holland, in order to procure a Patent there for our New invented Wind Engine or Machine I presume to give yor Lordship the trouble of these Lines; making my humble acknowledgments for the many Favours already received from you on that Ocasion, particularly for yor kind recomendation of us to a Person of so much Honour and Benignity as Sr Matthew Deyker, who was pleased not only to favour us with his own Letter to a very worthy Gentleman, but also to procure another for us from the Dutch Ambassador to the Great Man at the Hague, which had all the good Effect that could well be desired, and the Affair was like soon to be brought to a good Issue, had it not been prevented by the Decease of our Agent there: But whilst I am acknowledging past Favours, the Circumstances of our Affair obliges me to inform yor Lordship that we stand in need of further Assistance; for designing not to lose the fair Prospect we have of Sucess in it, the Persons concerned have agreed to send my Nephew Mr George Newcomen to Holland, in order to perfect that which hath been so happily begun; and supposing it will be very proper, if not necessary, that he should also be well recomended on his going thither, as the Person who is to succeed Mr Calley in the Prosecution of that Affair: If yor Lordship is of the same Opinion, I humbly hope & desire you would be pleased to favour us with a fresh Recomendation to Sr Matthew, who I perswade my self will be so good as to renew his former Favours and thereby compleat the Benefit he intended us, the full Effect of which hath hitherto been providentially prevented.

My Nephew, Mr Elias Newcomen, who is to wait on yor Lordship with this, will be ready to observe what Directions you shall be pleased to give him herein, and I shall allwayes be ready to acknowledge it as a great Adition to the many Obligations I already ly under, & with tenders of all due Respects am

Dartmo May 7th 1725

My Ld Yor Lordships very humble
and Obedient Servant

Tho: Newcomen

46 A letter written by Thomas Newcomen to Lord Chief Justice King from Dartmouth, dated 7 May 1725. The letter refers to the death of John Calley (Jnr), a new-invented wind engine and Newcomen's nephews, George and Elias Newcomen.

machine for raising water by fire'. Scant recognition, but at least it gave credit where it was due.

Lidstone, the Dartmouth historian, stated that he was buried in the nonconformist burial ground at Bunhill Fields, City Road, Finsbury. This was subsequently confirmed by a search in the Register Book of Burials which revealed the following entry dated 8 August 1729: 'Mr. Newcomen from St Mary Magdelans buried in a valt 00-14-00'. This would not be a marked grave but a common vault of which there is now no trace.

We have only a little information about his descendants. His elder son, Thomas, made a will in 1757. He had four children, Thomas, Hannah, Martha and William and he died in 1767 being buried at Dartmouth where the Church Register records him as an Anabaptist. It is coincidence that the inventor's son, Thomas, lived in the Parish of St Mary Magdalene, Taunton, and that his father was to die in a parish of the same name in London.

My Dear Wife London X^{ber} 30th 1727

I rejoyce to hear, by yo[rs] 26th Inst: that the ffamily is in good Health, which Mercy I am also favoured with.

I suppose Elias may be returned before this comes to Hand, if not remember me kindly to him, & to our other two Children, & tell them I should greatly rejoyce to hear they were seriously enquiring the Way to Sion with their ffaces thitherward: This ought to be their Chiefest Concern, as ever they propose to themselves the Enjoyment of True Happiness: Tell them that I sometimes reflect upon the Mellancholly Circumstance of the late Prince Menzikoff, who a few Months since was Prime Minister to the Great Emperour of Russia, had arrived to an Extraordinary Height of Power, had accumulated to himself an Imense Quantity of Riches, & was almost adored by all, as the most Happy of all Men, but was suddenly deprived of all, & reduced to his former Degree of Meaness, having also incurred the highest Displeasure of the Great Monarch; yet, in my Apprehension (notwithstanding the many sorrowfull Reflections he may be supposed to make upon it) his Case is very desireable, when set in Compare with that ffooll mentioned by our Saviour Luke 12th: Who, when his Soul comes to be required of him, shall be found only to have been laying up Treasure to himself & is not Rich towards God, for the former hath Time & Opportunity to provide himself of a far better & greater Treasure than what he hath lost, whilst the other is past all Hope in that Respect: The former hath Nothing more to fear from the Rage of his Great Master, than the Killing of his Body; But Oh! what hath not the latter to fear from the Anger of an Incensed God, who had so often Offered himself unto him as his Portion in order to his Everlasting Happiness, but was neglected & slighted; And for what was the Gracious Offer despised? Even for the Gratification of Sinfull Lusts, or for the Enjoyment of Lying Vanities, which he very well knew he must soon leave, and how soon who can tell? The Lord grant these Considerations may make suitable Impressions upon all our Minds, to his Care I heartily commend you & with dear Love to yourself & [illegible] Respects to all as due & [illegible] yo[r] [illegible] Happy [illegible] Company [illegible] to call the Meeting of that Week, so shall send the [illegible] & [illegible] next [illegible]. Mr Wallin being present gives her Service to you

Yo[r] Affect[ionat]e Husband

Tho: Newcomen

47 The only other letter written by Newcomen to survive. This was to his wife dated 30 December 1727 and was written from the house of Edward Wallin in London. The letter, almost in the form of a sermon, urges religious views on his children.

Newcomen's second son Elias, was recorded as an ironmonger of Clifton, Dartmouth Hardness and took over his father's premises and clearly followed the established family business. He married his first cousin Hannah Waymouth who was the eldest daughter of Samuel Waymouth of Exeter, a tobacconist. They had four children, Thomas, Samuel, John and Hannah. Samuel died at Charlestown, America in 1782 and John at Jamestown, America in 1805. Hannah was married to Nicholas Gibbs, a Linen Draper of Plymouth, who died in 1790 aged 34 and then to William Prance of Plymouth.

Elias became the administrator of Newcomen's affairs following the death of his widow and administratrix, so that strangely neither husband or wife left a will and we do not know what financial estate Newcomen left. Elias leased until his death the same properties in Dartmouth as his father had done. He died in 1765 and his wife went later to live in Plymouth with her daughter Hannah. Thomas Newcomen's daughter Hannah married William Wolcot of Clifton Dartmouth, a chirurgeon, or surgeon, but died without issue.

Although we now know that at least 100 engines were built before the expiry of the patent, the Proprietors seem to have been singularly reluctant to acknowledge Newcomen as the author of the great invention which they were exploiting. Before Newcomen died his engines were at work on many sites in Britain from the Firth of Forth to Mounts Bay and in Hungary, Belgium, France, Germany, Sweden, Austria and possibly Spain, but he received no honour in his lifetime. In Britain his engine was sometimes referred to as Savery's or, as an improvement on Savery's while abroad the laurels were awarded to whoever was responsible for building the engine concerned. Few of those who knew the true facts concerned themselves to correct these false attributions and so pay honour where honour was so justly due. Was Newcomen embittered by this lack of recognition? He seems to have maintained an amicable relationship with his fellow Proprietors to the end and the inference is that he was too modest a man to seek the world's acclaim and too great a man to be soured because he failed to receive it. To the creative genius, the work of creation is its own reward, a satisfaction which no one can take away and which armours the creator against the sharpest arrows of outrageous fortune. The life story of Newcomen's great successor, Richard Trevithick, illustrates this truth and it is fair to assume that Thomas Newcomen himself belonged to the same select company, to whom the work itself is all.

Fortunately recent interest in the story of Newcomen has ensured that his name has not been forgotten. In 1921 the Borough of Dartmouth set up a memorial comprising a stone base to which is affixed a brass plate with an engraving of his engine and an inscription. In Dartmouth too is the 'Memorial Engine', a small atmospheric engine donated by British Waterways and set up in 1963 by the Newcomen Society to mark the 300th anniversary of Newcomen's birth.

At the North Western Museum of Science and Industry at Manchester, Dr R. L. Hills has built a one-third scale working model of the 1712 engine and this is demonstrated to visitors. Even with today's technology many problems had to be overcome before the engine would work successfully and the setting of the valves was found to be most critical.

Perhaps the most exciting event is currently (1976) taking place on the site of the new Black Country Museum at Dudley in the West Midlands, where a full-sized working replica of the 1712 engine is being constructed. The site is very near to the Coneygree Colliery where the original engine was built and the project will demonstrate to the world the genius of the man and will at last form a worthy tribute to his immortal memory.

CHAPTER 5

Technical Developments 1712-33

In order to present the story of Newcomen and his engine as clearly as possible a great deal of technical detail has been omitted from the preceding chapters. Such detail is essential, however, to any assessment of Newcomen's achievement because so many conflicting statements have been made as to how, and by whom, certain features of the engine were developed. Quite apart from the need to pay Newcomen his due, the story of how men, who were quite ignorant of the nature of steam or the laws of thermodynamics, groped their way to success by sheer practical ingenuity and tenacity of purpose is a fascinating one.

To appreciate to the full the achievement of Newcomen and his associates, the reader must use his imagination, forget his twentieth-century background and try to think himself back into those far-off days. Such an imaginative effort is essential if the error of hindsight is to be avoided, because the solutions of many of the problems with which Newcomen grappled now seem so obvious as to be self-evident even to those who are not technically minded. Even some of Newcomen's contemporaries and immediate successors made this same mistake, so fatally easy to be wise after the event. The most brilliant inventions are those which seem obvious once they have been explained or demonstrated. 'Why, I could have thought of that', we say with a feeling of envy and mortification, and it is clear from their writings that both Desaguliers and Triewald reacted in this way, convincing themselves that they could have done better than Newcomen had they chosen to bend their minds to his problems.

Next to the principle of injecting water into the cylinder to create a vacuum, the most significant feature of the Newcomen engine was the valve gear, or 'working gear' as it was called, which made it self-acting. No early writer awards Newcomen any credit for this valve gear and the only legend about his engine that deserves to be called popular would have us believe that the valves were worked by hand until a boy, tiring of this monotonous task, connected their handles by strings to the beam. There may be a grain of misunderstood truth in this legend as we shall see presently.

Farey, in his *Treatise on the Steam Engine*, summarizes the evolution of the valve gear as follows:

> At first the valves were opened and shut by hand, and required the most exact and unremitting care of the attendant, to perform those operations at the precise moment; the least neglect or inadvertence might be ruinous to the machine, by beating out the bottom of the cylinder, or allowing the piston to be wholly drawn out of it. Stops were contrived to prevent both of these accidents; then strings were used, to connect the handles of the cocks with the lever [ie the balance beam] so that they should be turned whenever it reached certain positions. These strings were gradually changed and improved, into detents and catches of different shapes: till at last, in 1718, Mr. Beighton, a very ingenious and well-informed engineer, simplified the whole of these subordinate movements, and brought the machine into the form in which it has continued, to the present day, without any material change.

Stuart, in his *History of the Steam Engine*, goes further than this. His plate entitled 'Newcomen's Engine' depicts an engine with hand-operated valves, while the next figure showing an engine with self-acting valve gear is headed 'Beighton's Engine'. His accompanying text reads:

> The mechanism for opening and shutting the cocks also remained perplexed by catches and strings, until Mr. Henry Beighton, an engineer extensively employed in the construction of mining machinery, erected an engine at Newcastle-on-Tyne in 1718, in which all these "cock-boys" and complications of cords were superseded by a rod suspended from the beam, which operated on a mechanism invented by him called *hand-gear*: a contrivance, with some slight modifications, employed in engines of the present day.

Although it is not possible to back the statement with positive proof, it is safe to say that the hand-operated valves referred to by Farey and illustrated by Stuart belong to the experimental period before 1712. From that date forward Newcomen's engine had the self-acting valve gear evolved by him and his assistant Calley and differing only in one important respect from that which was later standardized and used with only minor variations down to the end of the eighteenth century. This one important difference accounts for the references to catches and strings as well as for the legends about the ingenious 'cock boy'. The fact that these references and legends are woefully misleading is due to a failure to understand the difficulties — some of them due to lack of knowledge and experience — encountered by Newcomen at the outset and the way he met them. Before these difficulties and Newcomen's solutions to them can be understood, it is essential to consider two things: the adequacy of the boilers used on Newcomen's first engines and this method of calculating the work these engines could perform.

It will be recalled that when Nicholas Ridley proposed to erect a second engine at his Byker colliery with a 33in cylinder the inventors, presumably Newcomen and Calley, refused to undertake the project on the grounds that so large a cylinder could not be supplied with sufficient steam. Commenting on this refusal, Triewald writes:

> The cause of this conclusion was the false principles concerning the steam which the inventors harboured in their minds according to which the steam rises or is generated by the boiling water in proportion to the quantity of water in the boiler. In consequence their boilers were made very high, as demonstrated by the Stafford [Dudley Castle] engine, the boiler of which is higher than its width. It is thus evident that the inventors do not know that the boiler must be given a suitable shape. Neither did they know that the flames should be allowed to play all around the side of the boiler as well as on the bottom. . . .

Although Triewald was a conceited young man and apt to be wise after the event, there is no doubt that in this case his criticism was valid. It may seem obvious to us that the steam generating capacity of a boiler depends on its heating surface but this was by no means obvious in 1712. To suppose that it was only necessary to increase the quantity of boiling water in order to obtain a proportionate increase in the volume of steam produced from it was then a perfectly understandable error. Moreover, although Triewald appreciated the need to increase the heating surface, this conclusion was purely empirical and boilers continued to be built on an empirical basis for more than sixty years. The first Watt engines were under-boilered and this defect led Watt to work out, for the first time, a desirable ratio between boiler heating surface and cylinder volume. He specified four square feet of heating surface per cubic foot of cylinder volume.

The type of boiler used by Newcomen consisted of a cylindrical copper vessel with a concave bottom directly above the furnace. This was surmounted by a hemispherical steam dome of lead. The diameter of this dome was greater than that of the copper cylinder below, the upper portion of the latter being flanged outwards at right-angles and then turned upwards again in order to form a circumferential seam with the lead dome. It thus resembled a haystack in cross-section and it became know as a ' haystack' or 'flange' boiler, sometimes also called a 'balloon' or 'beehive' boiler. This was certainly the type of boiler supplied initially to the Whitehaven engine, although Spedding stated that it was the first of its type that Newcomen and Calley ever had and they did not approve of them as much as those without flanges. The brickwork enclosing the furnace was carried upwards round the boiler and sealed against the lead dome plates. This arrangement left an annular space beneath the dome and through this the hot gases from the furnace circulated before passing up the chimney stack. The proportions of the boiler were such that at working level there was but little depth of water upon the copper 'roof' of the annular flue and this made it vulnerable in the event of mismanagement. Although at an early date, high and low level try-cocks were provided so that the water level could be checked, attendants all too frequently allowed the level to fall low with the result that the copper 'roof' overheated and this led to the failure of the copper-lead seam at its circumference. The Austhorpe engine is said to have burned out four boilers in this way in as many years. The first boiler at Whitehaven suffered the same fate and had to be repaired first with lead and copper patches and then lined all round with lead held on with lead nails. Although the eventual substitution of wrought-iron plates for copper and lead made the boiler less liable to suffer from overheating, what was known as the 'tun' boiler came to be favoured by later builders. Instead of the angular construction of the Newcomen 'haystack', the lower portion of the tun boiler was cylindrical, tapering down from the diameter of the steam dome to that of the concave bottom plate immediately above the furnace. This entailed a certain loss of heating surface, but this sacrifice was offset by altering the proportions of the boiler and its furnace and flues. The second boiler at Whitehaven was made from iron plates with a lead top and supplied by Stonier Parrott in 1717. It was reported as being broader at the bottom and narrower at the top than the first boiler. This was to give trouble in service at the junction of the iron and lead and the iron corroded externally in this area. As early as March 1717 Mr Calley had commented that 'he approved very well of iron boilers' indicating their use elsewhere prior to this time.

So ingrained in us is the importance of a large heating surface that modern diagrammatic representations of the early Newcomen engine almost invariably depict a haystack boiler of a diameter exceeding its height. By contrast, Barney's engraving of the Dudley Castle engine shows a brick furnace and boiler casing of considerable height in proportion to its diameter. It is particularly interesting to note that Barney shows the firehole door in an impossible position above an arched access doorway to the ashpit so high that the fireman depicted in the foreground could walk through it without stooping. Without the aid of a stepladder he would have had to be a gymnast to put coal on the fire and would certainly need the 'little Bench with a Bass to rest when they are weary' which, according to Barney's key, was thoughtfully provided. This odd aspect of the drawing may be due to Barney's method of showing the ash hole and engine-water pump below the engine-house floor. The engine man is standing at the firing-floor level but this part of the drawing is moved forward to permit the lower level to be seen. Barney's accompanying key tells us that the boiler was 6ft 1in high but that the diameter of the bottom plate was only 4ft 4in. Barney also gives the capacity of the boiler as 'near 13 Hogsheads', this being the equivalent of approximately 680 imperial gallons. Having regard to the small heating surface available, this was an immense volume of water and Barney's figures thus corroborate Triewald's criticism of the Dudley Castle engine. It is clear that, as he says, Newcomen did believe at this time that if he increased the volume of water in a boiler its output of steam would increase proportionately. The full significance of this will appear presently.

It is also interesting to note that the von Schönström drawings of the Königsberg engine show a plain boiler without flanges.

The valve or 'regulator' that controlled the admission of steam from the boiler to the bottom of the working cylinder was mounted in the top of the steam dome and it resembled the type of valve used by Savery on his later engines. It consisted of a fan-shaped brass plate which, moving to and fro horizontally, alternately covered and un-covered a steam port formed in a second brass plate rivetted into the top of the boiler. On some engines the moving plate was held to its port face by a flat steel spring, bridging the steam orifice below and bearing against a boss formed on the lower surface of the plate. From the fixed brass plate in the top of the boiler the steam pipe extended upwards to meet in a butt joint a pipe of the same diameter passing through the centre of the cylinder bottom. This butt joint was wrapped, first with canvas covered in white lead and oil, then with sheet lead and finally bound tightly with cord. This primitive method of jointing persisted for years despite frequent failures caused by movement of the cylinder while the engine was working under load.

The vertical spindle of the moving sector valve was tapered like the barrel of a plug cock where it passed through the fixed brass plate in the boiler top and it was ground into a taper seating formed in the latter in order to keep it steam-tight. To the squared end of this spindle a spanner was attached which could be worked to and fro by a linkage arranged as follows. First there was pin-jointed to the spanner a horizontal link called the 'stirrup' because it was so shaped at its opposite end. The forked ends of this stirrup were supported by two vertical links, hanging loosely from a horizontal arbor which they shared with two levers called the 'Y' lever and the 'little Y'. Both were fixed at their common axis. The Y lever was in the shape of that letter inverted and it carried a weight or tumbling

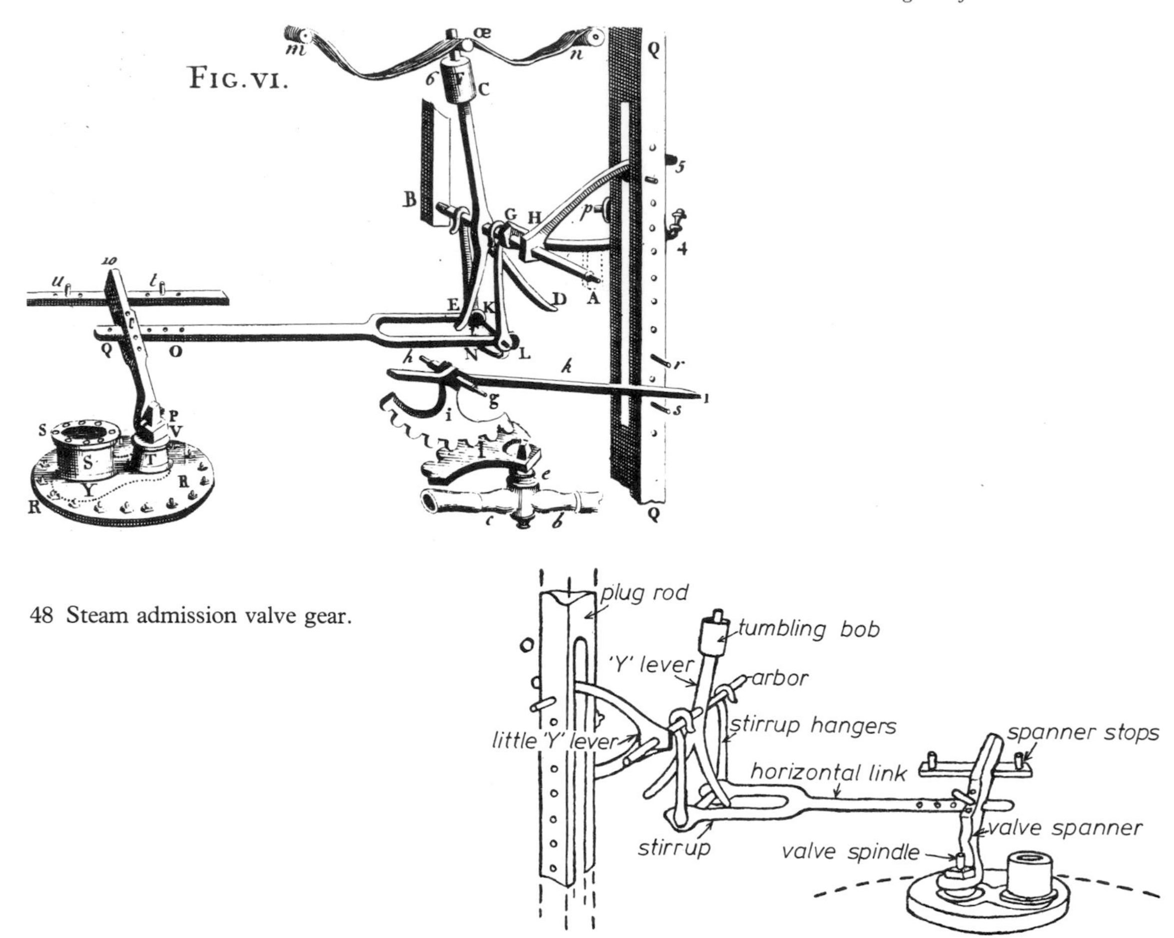

48 Steam admission valve gear.

bob on its upper end. As it rocked on its arbor and the weight was thrown over centre, its two lower arms alternately struck the crosspiece of the stirrup and thus opened and closed the valve, the action of the weight ensuring rapid and positive action. The plug rod which hung from the beam imparted this rocking motion to the arbor by means of the little Y lever. Pegs in the plug rod alternately raised and depressed the arms of this little Y as the rod rose and fell. The valve spanner itself was provided with positive stops and there were two methods of adjusting the motion. The pegs in the plug rod could be moved into alternative holes, while a series of holes in the arm of the stirrup enabled the pin-joint by which it moved the spanner to be adjusted similarly. This type of steam valve and its operating linkage persisted almost unchanged for many years with only detail refinements and it is safe to credit its inception, not to Henry Beighton, but to Newcomen and Calley. The design of the valve itself may have been derived from Savery, but the ingenious self-acting gear was entirely original.

The admission of injection water to the cylinder was controlled by a simple plug cock. To the designer of the steam admission gear it would have been a simple matter to develop a similar linkage to actuate this cock from the plug rod, but Newcomen did not do this for two closely associated reasons: the limited steam generating capacity of his boiler and the excessive load he imposed on the engine.

Desaguliers, who was doubtless supplied with the information by Henry Beighton, has this to say about Newcomen's method of estimating the power of his engines:

Mr *Newcomen's* Way of finding it was this: From the diameter [of the cylinder] squar'd he cut off the last Figure, calling the Figure on the left Hand long Hundreds, and writing a Cypher on the right Hand, call'd the Number on that Side, Pounds; and this he reckon'd pretty exact as a Mean, or rather when the Barometer stood at 30 and the Air was heavy. N.B. *This makes between 11 and 12 pounds upon every superficial round Inch.* Then he allow'd between 1/3 and 1/4 Part for what is lost in the Friction of the several Parts and for Accidents: and this will agree pretty well with the work at *Griff* Engine, there being lifted at every stroke between 2/3 and 3/4 of the weight of the atmospherical Column pressing on the Piston.

This works out at a mean effective pressure of 9·4 lb/in², an optimistic figure bearing in mind that the atmospheric pressure employed could not exceed 14 lb/in². Seventy years later the cautious Watt, although he was using steam at a pressure slightly exceeding that of the atmosphere, allowed a mean effective pressure of only 5 lb/in² on his first rotative engines. The figure of 9·4 lb/in² adopted by Newcomen meant that his first engines were so loaded that their working depended on the creation of a high degree of vacuum below the piston.

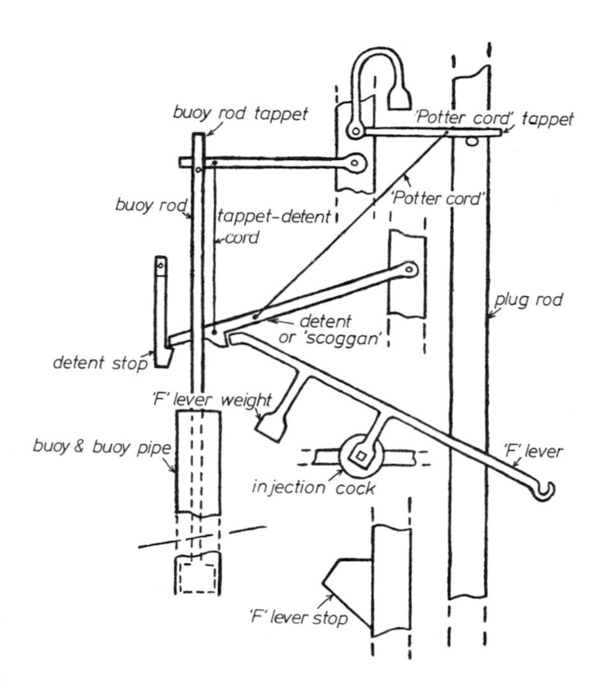

49 Injection valve gear.

When the engine had made its power stroke, or 'come into the house' to use the current engineman's expression, the piston was returned to the top of the cylinder by the weight of the descending pump rods at the other end of the beam. So far from being asssisted in this motion by the incoming steam, the pressure of the steam above atmosphere being negligible, the piston in its ascent actually helped to draw the steam from the boiler into the cylinders. The steam-generating capacity of the first boilers was so inadequate that the sudden extraction of so great a volume of steam might literally have the effect of sending the boiler 'off the boil'. If, therefore, water was immediately injected into the steam-filled cylinder so that the engine made another power stroke, the piston would once again be returned by the pump rods but little or no steam would follow it because the boiler had not had time to recoup. This, of course, would bring the engine to rest because in the absence of steam no vacuum could be created. Faced with this problem Newcomen decided that the solution was to regulate the working capacity of the engine in accordance with the steaming capacity of the boiler.

This solution took the form of a contrivance called a buoy. This was a small buoy floating upon the surface of the water in the boiler and enclosed in a vertical tube — the 'buoy pipe' — which protruded through the dome of the boiler and carried within it a rod attached to the buoy. By means of this rod the buoy controlled the opening of the injection cock in the following way. This cock carried a shaped lever which became known as the 'F' because of its shape. With the cock in the closed position, this F-lever lay at an angle of about 30° to the horizontal, with its foot nearest to the plug rod and the letter inclining upwards therefrom with the two short arms pointing downwards. The shorter of these two arms was pierced with the square hole fitting over the spindle of the injection cock, while the longer carried a weight at its extremity. A short prolongation of the lever beyond the junction of this upper arm engaged with a pivoted catch or detent lever known as the 'scoggan'. This detent retained the lever in its closed position against the reaction of the weighted arm which would otherwise fall and cause the injection cock to open.

Towards the end of the engine's power stroke a peg on the descending plug rod depressed the foot of the F-lever until it reached the closed position where it was retained by the detent. It was at this point in the cycle that the buoy came into play. The steam admission valve having opened and refilled the cylinder with steam as the piston 'went out of the house', the buoy ensured that another power stroke was not made until the boiler had recouped itself sufficiently to cope with the next demand for steam which must follow the power stroke immediately. When the dome of the boiler had again filled with steam, the slight pressure so created was sufficient to cause the buoy to rise in its pipe, whereupon the vertical rod attached to it raised the detent lever, thus enabling the F lever to fall by the action of its weight and so open the injection cock. It is important to note that in devising this ingenious cycle of operations Newcomen took it for granted that the steam demand of the cylinder would temporarily exhaust the boiler and so allow the buoy to fall. For if the buoy did not fall it would prevent the detent lever from returning to position in order to retain the F lever when the descending plug rod again restored the latter to the closed position.

It follows logically that so long as the engine was controlled by the buoy its action would be extremely slow. Whereas the word 'engine' to us implies continuous motion, on the first Newcomens each return of the pump rods would be followed by a prolonged pause while the boiler regained the necessary strength to perpetuate the cycle of operations. It seems clear, however, that, partly as a result of this extremely slow action, Newcomen did achieve a high degree of vacuum in his cylinder which enabled the engine to work a load as high as 9·4 lb/in^2 on the piston. This compensated to some extent for the very slow rate of pumping.

Notwithstanding the fact that instead of positive evidence we have a fog of confusing and contradictory statements, the subsequent development of the engine can be deduced with the assurance that, although it is necessarily conjectural, it cannot be far from the truth. One explanation of the legend of the ingenious boy with the piece of string is that it arose as a result of a misunderstanding of the fact that the 'buoy' used by Newcomen was not of the two-legged species. Although such a misunderstanding may have added to the confusion, it seems far more probable that such a boy did exist but that those who mentioned him did not understand what he was set to do or what it was that he achieved with his piece of string.

It is obvious that, as soon as a Newcomen engine was provided with a boiler of adequate steam-generating capacity, it became no longer necessary for the boiler to govern the working rate. Moreover, the old control gear could no longer operate because the buoy would keep the detent or scoggan permanently raised so that there would be nothing to retain the injection cock in the closed position during steam admission. When this first occurred there was only one way in which the engine concerned could be kept at work. This was by putting the buoy out of action by wedging its rod in the pipe and then lifting the detent by hand each time the piston reached the top of its stroke. The engine would then work at a much faster rate provided the boiler could continue to maintain the steam supply. To perform this monotonous operation, a two-legged boy was pressed into service instead of Newcomen's similarly named device. Standing close beside the rising and falling plug rod, it would not take this boy long to realise that the plug rod could very easily be made to do his repetitive job for him. A suitably positioned nail in the plug rod and a length of cord from the nail to the detent lever were all that was necessary. How delighted he must have been with his crude but effective improvisation as each time the plug rod neared the top of its stroke the cord tautened, raised the detent and so allowed the injection cock to open!

John O'Kelly tells us of the boy's invention:

> at the beginning, they only made 6 to 8 and 10 strokes per minute and it was as a result of the invention of a youth who watched over the machine that they managed 15 and 16 strokes in the same period of time. This boy was called Humphrey Potter, but this invention made the machine very complicated.

Young Humphrey Potter was most probably the brother of Isaac and John who were the sons of Stephen Potter, brother to Humprey, Sen, and clearly a family trio who had a tremendous influence on the introduction and development of the engine.

Desaguliers however gives a most muddled account of the affair as follows:

> They used to work with a buoy in the cylinder, enclosed in a pipe: which buoy rose when the steam was strong, and opened the injection and made a stroke: thereby they were only able from this imperfect mechanism to make six, or eight, or ten strokes in a minute; till a boy named Humphry Potter, who attended the engine, added what he called a *scoggan* – a catch, that the beam (or lever) always opened; and then it would go fifteen or sixteen strokes in a minute.

It is evident from this that the worthy scientist totally failed to grasp both the object and the working principle of Newcomen's buoy gear. He begins by placing the buoy in the cylinder, then blames the 'imperfect mechanism' for the fact that the engine worked so slowly and finally credits the boy Potter with the introduction of the detent lever or scoggan; all this to the infinite bemusement of subsequent writers.

A good example of the way in which nonsense begets nonsense appeared long afterwards in the *Mechanic's Magazine* and was quoted by Stuart. Accepting Desaguliers's account, the writer announced that the word scoggan was derived from a north Yorkshire verb 'to scog', meaning to skulk or idle and that it was obviously applied to the detent lever because it enabled its youthful inventor to idle instead of attending to his monotonous job. In fact the word is of Cornish origin and was doubtless first applied to the detent lever by Newcomen or his associates during the course of pre-1712 experiments in Cornwall. The term is still applied in Cornwall to the Cornish engine valve gear, but it will soon become an archaism now that this ultimate development of the non-rotative beam engine has become a museum piece like its predecessors.

It should be emphasized at this point that in designing his sorely misunderstood method of operating his injection cock. Newcomen was the first to adopt the principle of opening the valve by the fall of a weight and closing it against gravity, a principle which has proved as long-lived as the word scoggan, since it persisted through the Watt era down to the last Cornish engine to be built.

The acceleration of the engine from a mere six strokes a minute to twelve of fifteen strokes by the use of a more efficient boiler and the substitution of the 'Potter cord' for the buoy to regulate the opening of the injection valve was not achieved without loss in another direction. The faster rate of working allowed less time for the cylinder alternately to gain and lose its heat with the result that the degree of vacuum achieved below the piston was reduced. The mean effective pressure of 9·4 lb/in^2 originally adopted by Newcomen was reduced to 7 lb/in^2 or in other words half atmospheric pressure, to counteract this loss, but it was immediately found that the less heavily loaded engine working at the faster rate pumped more water per hour than it did before. Nor was any more fuel consumed in proportion to the work done because the more efficient boiler wasted less of the heat generated by the furnace.

If contemporary engravings are to be relied upon, it would appear that the earliest Newcomen engines were soon fitted with a refined version of the 'Potter cord' but that the buoy gear was retained. Beighton (1717), Barney (1719), and, most surprisingly, Triewald's engraving of the Dannemora engine (1734) all show both. No doubt when the boiler was steaming badly and could no longer sustain the rate of working imposed by the Potter cord, it was disconnected so that the buoy could take control. Meanwhile, however, the provision of more adequate boilers was accompanied by the rapid accumulation of experience among the men concerned with the erection and management of the engines with the effect that the time soon came when it was felt that the buoy gear could be safely discarded. By far the most likely explanation of the claim that it was Beighton who 'invented' the self-acting valve gear in 1718 is that it was he who first took this step by dispensing with both the buoy gear and the Potter cord on the engine which he erected at Oxclose Colliery, Washington Fell. This would also explain the statement that until this time the engine had remained 'perplexed by catches and strings'.

Beighton is also said to have been the first to fit a weighted safety valve, or 'puppet clack' as it was called, on the boiler of this engine at the suggestion of Desaguliers. It lifted at a pressure of $1\frac{1}{2}$ lb/in^2. This sounds logical since the elimination of the buoy would make a safety valve the more necessary, but the fact remains that a weight safety valve features in Beighton's engraving of 1717 along with the buoy and the Potter cord.

Stuart's engraving of what he calls 'Beighton's Hand-gear' shows a very simple method of working the injection cock by means of two toothed sectors engaging in each other at right-angles, the driving sector being mounted on the horizontal axis of a lever which was alternately moved up and down by two pegs in the plug rod. One of the early engravings featuring a gear similar to this is that showing the engine at Passy (1726) which illustrates the French account of that engine. If this was Beighton's gear, it was soon discarded in favour of a return to the weighted lever released by a detent. Desaguliers describes and illustrates the toothed sector method of opening the injection cock, but also illustrates the weighted lever and detent, which he says is 'more used, and I think a great deal better; because it moves with a Jerk, which is the best way to overcome Friction'. The method by which this lever turned the cock might vary and, instead of the Potter cord, the detent was henceforth tripped by a striker on the plug rod, but the principle of opening by falling weight became firmly established. Experience has shown that the more rapidly and positively the injection cock could be opened, the better the result and it was for this reason that the simple gear credited to Beighton failed to supplant the weighted lever and detent first developed by Newcomen.

It must be emphasised that early pictures of Newcomen engines are not an infallible guide to chronological development. Artists either failed to understand the principle of the valve gear and drew it indistinctly or inaccurately, or else they copied their predecessor's work. Thus the Sutton Nicholls engraving of what purports to be the York Buildings engine (1725) shows the buoy gear only, which is certainly incorrect, while it seems most unlikely that the Dannemora engine would have had the buoy gear unless it was that Triewald doubted the ability of his boiler to supply so large a cylinder. Such vagaries become excusable when we realise that with the sole exception of clock-work, no other self-acting mechanism existed in those days.

Besides the alternative admission of steam and injection water to the cylinder, provision had to be made for the exhaustion from the cylinder of the hot condensate and of any air brought in by the steam. When steam was first admitted, the hot water, which had accumulated in the cylinder bottom during the previous stroke, was expelled down a pipe into a hot well. A leather non-return valve at the foot of this eduction pipe prevented the water being drawn back up the pipe when a vacuum was created in the cylinder. In the earliest engines the hot well was at ash-pit level, but later it was located above the boiler steam dome so that the hot water it contained could be fed by gravity into the boiler. Beighton is said to have been responsible for this innovation. This first form of feed water heating brought about a significant saving in fuel.

Apparently Newcomen did not at first appreciate that a certain amount of air would be carried into the cylinder with the steam and he was accordingly mystified by the fact that his engine would gradually lose power until it finally stopped, air having accumulated to such an extent that an effective vacuum could no longer be created. When the cause of this malady, which became known to enginemen as 'wind-logging', was recognised, Newcomen cured it by fitting a small outlet pipe to the lower part of the cylinder through which the incoming steam could expel any air. Like the eduction pipe, and for the same reason, it was fitted with a non-return valve. Because of the noise it made, this air outlet pipe became known as the snifting valve, 'snift' being then the equivalent of our 'sniff'. The pipe was led into a small tank of water which Barney calls a 'snifting bason'. There any steam which passed was condensed while the air bubbled through it. The overflow from this tank, led either to the hot well or directly to the boiler.

A cistern, mounted a little below the axis of the beam in order to provide a head, supplied the water for injection. It was replenished by a small pump worked, like the plug rod which operated the valve gear, by a small auxilliary arch-head set closer to the axis of the beam than the main arch-heads so that the stroke was reduced. In Beighton's engraving the arch-head for this pump is shown outside the house near the pump end of the beam. Barney's engraving, on the other hand, shows the pump in the engine house and driven from the end of the plug rod. The pumps used for this purpose were usually of the type known as 'jack-head'. They were of the common lifting type, but the top of the working barrel was closed, the bucket rod passing through a leather-packed gland in the cover. As the bucket ascended, the water above it was forced up a pipe which branched from the top of the working barrel. If Barney's drawing is correct, however, the Dudley Castle engine was fitted with a plunger force pump of the type introduced by Morland. This pumped on the downstroke and to cope with this the plug rod is shown coupled to the beam by two opposed chains anchored to the midpoint of the arch-head. Such an arrangement cannot have worked satisfactorily and a jack-head pump driven by a second auxilliary arch-head on the pump end of the beam became the rule.

The water seal on the top of the piston was replenished by a branch taken from the injection-water supply pipe. Barney's drawing omits this supply pipe and shows both the water seal and injection pipes branching from the pump delivery pipe which delivers into the top of the cistern, an arrangement that could not work satisfactorily and makes the cistern a mere ornament. The top of the cylinder was belled out to prevent the water on the top of the piston from spilling over at the top of the stroke. On the first engines the surplus water from the top of the cylinder was led from the bell-mouth by an overflow pipe directly back to the boiler. The temperature of this water was never very high, however, and it was led to waste after this arrangement had been abandoned in favour of drawing feed-water from the hot well.

50 A Newcomen engine piston.

There was no means of machining the brass cylinder internally: its bore had to be laboriously fettled and smoothed by hand. The piston was usually of cast-iron and, according to Desaguliers, Newcomen first used as a seal a disc of leather above the piston with its periphery upturned so that it became, in effect, a gigantic cup washer. This leather very speedily wore away in such a manner that the upturned portion broke away to leave only edge contact between the leather and the cylinder wall. Desaguliers goes on to say that Newcomen was delighted to find that this made an effective seal. The use of such seals is confirmed in the description of the engraving by Sutton Nicholls of 1725. The arrangement adopted and standardized on some of the early commercial engines was to cast the piston with an annular flange on its upper side, the outside diameter of this flange being three inches less than that of the piston head below which fitted the cylinder as closely as the techniques of the day permitted. With the piston in the cylinder, soft hemp packing was then coiled and rammed into the space between the face of the flange and the cylinder wall and finally segmental weights were added to hold this hemp packing tight and in place. The function of the water seal was to make this packing effective by keeping the hemp soft and pliable. It could not, as is sometimes supposed, seal a seriously defective or irregular cylinder bore because in such circumstances so much water would pass the piston that creation of a vacuum would be seriously impaired or prevented by excessive condensation of steam on admission.

The beam or 'Great Lever' consisted of either a single massive oak timber or of two such timbers secured together. The timber arch-heads were

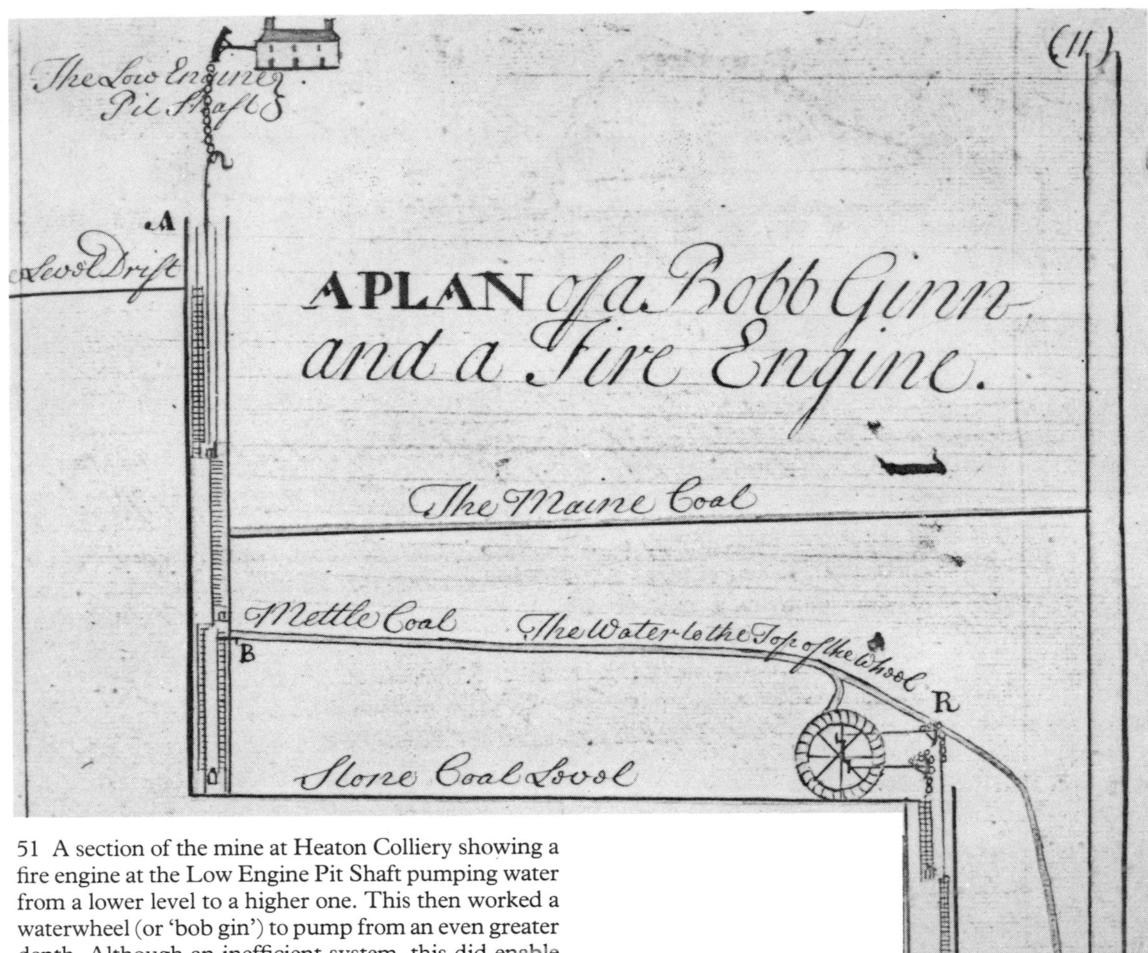

51 A section of the mine at Heaton Colliery showing a fire engine at the Low Engine Pit Shaft pumping water from a lower level to a higher one. This then worked a waterwheel (or 'bob gin') to pump from an even greater depth. Although an inefficient system, this did enable parts of the mine at a distance from the shaft foot to be kept dry.

mortised into the ends of the beam and securely braced above and below by timber diagonals. The upper-portions of the arch-heads were additionally braced by a stout iron rod passing right through the arch-head and so anchoring the chain by which piston and pump rods were suspended. In all early engines, the trunnions were placed at the mid point of the beam so that the strokes of piston and pump bucket were the same.

Some latter-day observers have wondered why, throughout the long history of the beam pumping engine, the pump end of the beam and its attendant gear should have been exposed to the elements outside the house. Newcomen rightly decided that the load on the beam trunnion bearings was such that they must be supported by the main wall of the engine house. It was called the lever wall and was more massively built. To enclose the whole would thus have involved the construction of additional walls and roofing which would have proved obstructive when it became necessary to withdraw the pump rods from the mine shaft. Very often these early mine shafts were used for access into the workings and for raising coal and minerals as well as pumping, so it was clearly impractical to enclose the top of the shaft with an engine house. Examples of totally enclosed engines with beam trunnion bearings supported on massive columns or 'A' frames may be seen, particularly in waterworks installations, but in the case of the Cornish mine pumping-engine Newcomen's practice of using wall support persisted right down to the early years of this century when the last Cornish engines were built.

On a non-rotative beam engine the length of stroke is not positively determined as it is by the crank of any type of rotative steam engine. When the engine was started by the hand manipulation of the levers controlling the steam and injection valves, the length of stroke it made depended on the skill of the engineman. When the self-acting gear took over control, the length of stroke was still determined by his judgement because successful transition from manual to automatic control depended on the correct placing of the pegs in the plug rod which actuated the valve levers. It was therefore neccessary to provide a form of positive stop to prevent the piston coming out of the top of the cylinder or, through mismanagement, knocking the bottom out of the cylinder on its descent.

When first admitted to the cylinder, the steam might exert a small power on the piston, but this diminished rapidly as the piston ascended until, as it neared the top of the stroke, it was literally drawing the steam out of the boiler. But when a new engine had been completed it had to be most carefully balanced to prevent the great weight of the descending pump rods drawing the piston up with a violence which would damage both the engine and the pumps. Weights were added to the beam immediately behind the piston arch-head and to the piston itself until the beam was in perfect equipoise. A weight equivalent to 1 lb/in^2, of piston area was then removed from the piston to give the pump end of the beam the necessary advantage. Naturally, if the mine was deepened and additional pump rods were added the engine would have to be rebalanced. In calculating the correct balance in favour of the pump end, account was taken of the resistance of the pump

52 William Pryce in *Mineralogia Cornubiensis*, 1778, includes a drawing of a Fire Engine as used in Cornwall. The drawing shows the use of a balance bob to counterweight the weight of the long pump rods.

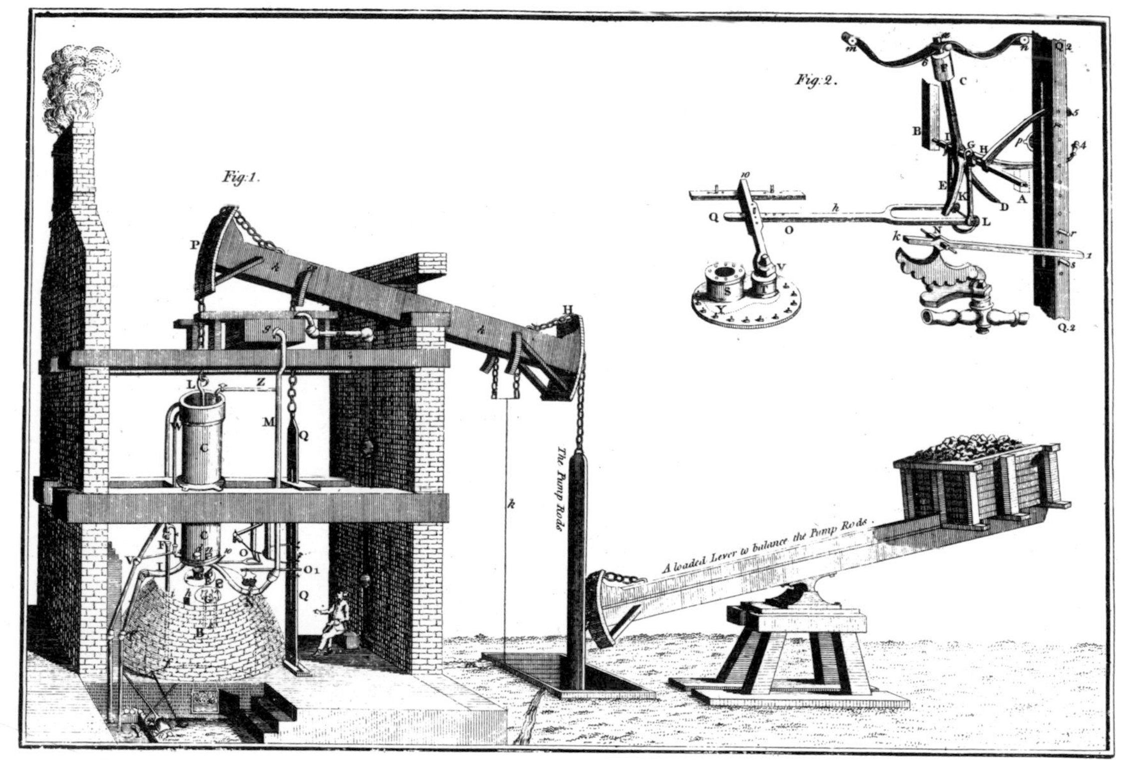

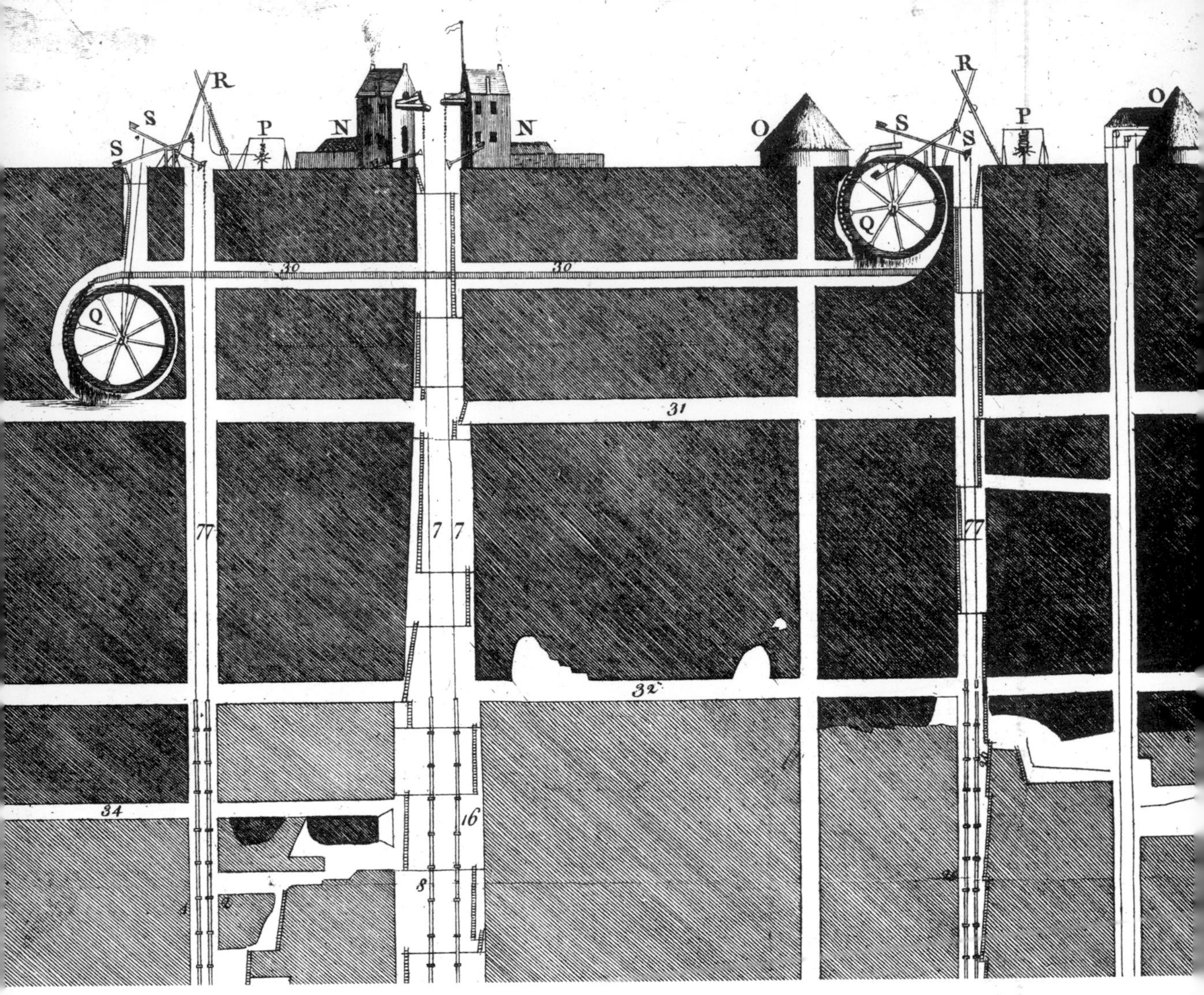

53 Pryce includes this section of the Bullen Garden Mine, Camborne, Cornwall. It indicates both the use of two engines with balance bobs and two waterwheels draining higher levels of the mine.

bucket as it descended through the water in the pump barrel, the passage of the water past the bucket being restricted by the orifice of the non-return valve in the bucket. The danger here occurred if the level of water in the mine sump was so reduced that the pump drew air instead of water. In that event the resistance of the air below the bucket might be insufficient to open the non-return valve when it began its descent. If this happened the weight of the whole column of water above the bucket would drive it down with great and damaging violence before compression of the air became sufficient to take effect. To guard against this the engineman was provided with a float-operated indicator which recorded the depth of water in the mine sump. Even so, some form of positive stop was essential to arrest the descent of the pump bucket and this took the form of sprung timbers known as 'spring beams' which provided a limited cushioning effect before becoming a positive stop. The pictures of the Dudley Castle and Dannemora engines show such spring beams mounted on stout timber stages above the mine shaft. The moving stop connecting with them consisted of a strong iron rod which passed through the top of the arch-head and projected on each side of it. This became known as the 'sword'. The alternative, which soon superseded this arrangement, was to mount the spring beams on a staging within the upper part of the mine shaft, the sword being then carried by the first length of pump rod. Presumably this arrangement was used in the engine portrayed by Beighton since he shows no overhead spring beam staging on the outdoor end.

Inside the engine house a precisely similar arrangement was used to prevent the piston from damaging the cylinder bottom if it 'came into the house' too violently. In this case the spring beam stagings were carried on two beams extending from wall to wall of the house as shown by Barney and Beighton. The artist of the Dannemora engine has omitted them in error because the supporting beams are shown. Below these supporting beams two even more massive timbers bridged the engine house in order to carry the cylinder, the latter having a cast flange which rested upon them.

Most early pictures of Newcomen engines show single forged-link chains coupling piston and pump rods to the arch-heads. These were soon discarded in favour of pin-jointed chains with flat plate links which were easier to repair or renew, lay more snugly on the arch-heads and were more readily coupled to cross-heads when two or more sets of pump rods were used. Duplex or multiple chains took the place of single chains on both arch-heads as mines deepened and engines increased in power.

54 The use of quadrants and horizontal connecting rods is shown in this drawing of 'The Slide Engine at Mill Close 1756', one of the London Lead Company's lead mines in Derbyshire. This is an early example of the use of flat rods, widely used in Cornwall in the next century, to pump from a shaft some distance from the engine.

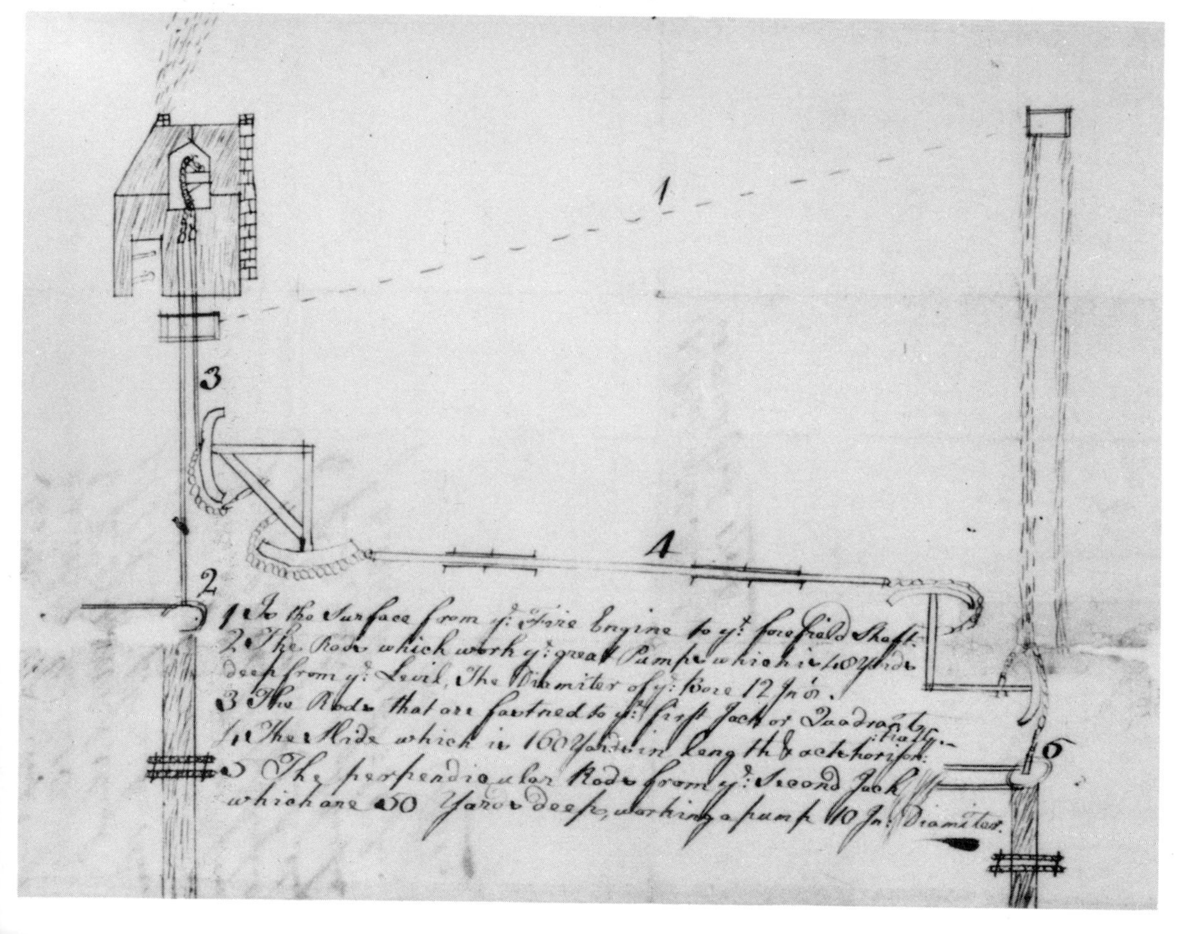

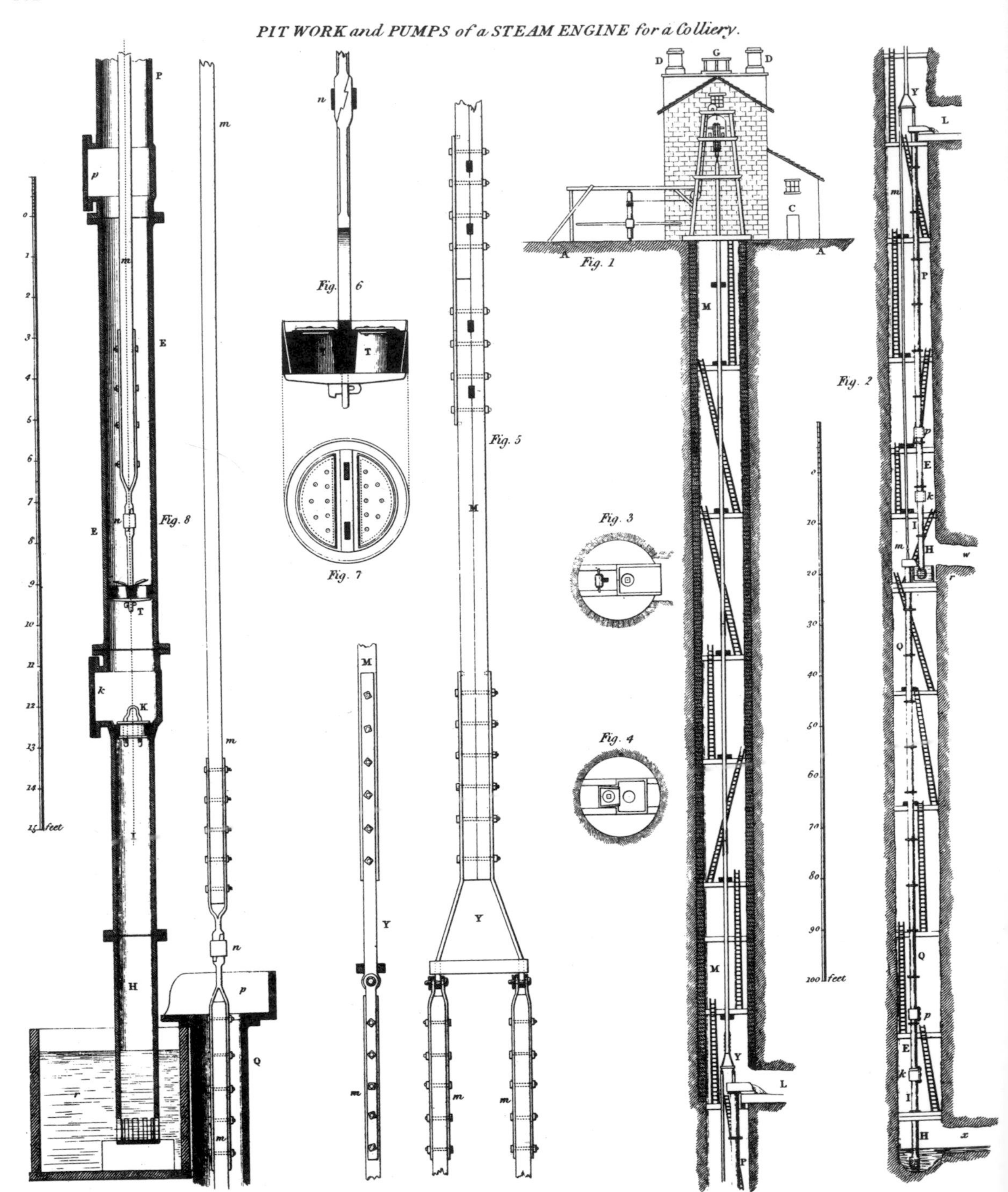

55 The pitwork and pumps of the Long Benton Colliery Engine.

Little has been said so far about the arrangement and construction of the pumps in the mine. It seems most unlikely that Newcomen was responsible for any notable innovation in this department, but his engine, by bringing much more power to bear, necessitated the rapid development of techniques which were first applied in mines a little earlier. The first mechanical means of raising water from mines consisted of chain-and-bucket pumps powered by waterwheels, or by horse gins in situations where water power could not be harnessed. Rod-operated lifting pumps began to supersede bucket pumps in the first decade of the eighteenth century, such pumps being driven by crank from a waterwheel. George Sorocold, an English engineer noted for his water-powered water supply installations, is said to have been responsible for introducing lifting pumps to the Scottish collieries in 1710 and their use soon spread to Tyneside through the agency of the Earl of Mar who owned property in both districts. It seems probable, too, that lifting pumps were beginning to supersede chains and buckets in Cornwall at the time when Newcomen was making his first experiments there.

While the means available were so inadequate, miners were prepared to go to immense pains to avoid the necessity of pumping. Adit tunnels were driven to carry the water away to low ground and in some districts, notably in Derbyshire, these tunnels were often of great length. The introduction of Newcomen's engine enabled mines to be worked below the level of these adits, but they were still used since it was obviously uneconomic to raise water farther than was absolutely necessary. If the adit level was far below the top of the mine shaft, it meant that the pump end of the engine beam had to carry a great length and weight of dry rods, or 'dry spears' as they were called, which necessitated the use of very heavy counterweights. To avoid loading the beam to such an extent, the practice of using an auxilliary beam or 'balance bob' which was mounted at the mouth of the shaft or as a separate beam above the engine main beam, was initiated. One end of this beam was coupled to the pump rod, while the other end carried a box filled with blocks of stone or scrap metal to serve as a counterweight to the rods. This not only relieved the load on the beam but greatly simplified the task of rebalancing when alterations in the mine made this necessary.

In order to save weight the pump rods or 'spears' were of fir or mast timber in lengths of from 40 to 60ft, the joints between them being scarfed and held together by crossbolts and fishplates. In the case of the 'wet spears' that passed down the pipe through which the water was drawn, the buoyancy of the wood in the water relieved the weight on the engine. To provide

56 Perspective view of a mine shaft showing pumps. Thos Martin *Cycle of Mechanical Arts*, 1813.

57 A view of Chelsea Water Works in 1752. Close scrutiny of the original reveals that both engines have two chains hanging from the arch heads and would have worked two pumps. On the right of the engines is a house crane for unloading the coal from the sailing barges. The building in front of the crane is perhaps a coal store. Timber water pipes lie in front of this store.

access to the mine there was a series of timber stages interconnected by ladders in the mine shaft and on each of these stages the pump rods were guided by rollers to prevent them from bowing as they descended.

Forty-five yards was considered the maximum lift for a single pump and in practice the single lift was usually substantially less than this. In deep mines this necessitated a series of pumps placed at different stages down the shaft, each drawing from a wooden cistern filled by the pump next below. Such pumps would be driven by rods of smaller section coupled by crossheads to the main rod. Such a series of pumps would be identical except that the lowest, which drew from the mine sump, would have a different suction pipe or 'wind bore'.

As we know from the Edmonstone and Whitehaven engine accounts, the earliest pipes through which the water was lifted were of elm banded with iron, but these wooden pipes were soon superseded by iron pipes with flange joints. The lowest length of pipe next to the working barrel of the pump was of slightly larger diameter than the rest and was provided with an access door. Into this pipe the bucket of the pump could be drawn when it, or the valve within it, required repair or replacement and on this account it was called the 'bucket piece'. Below this, the working barrel of the pump was generally of brass or bell metal to resist corrosion and would be 9ft long if the working stroke was 7ft. Immediately below the pump barrel was a length of pipe called the 'clack piece'. Its expanded upper end formed a conical seating for the clack or foot valve, an access door being provided similar to that in the bucket piece. The foot valve was fitted with an iron loop which could be grappled from above after the pump bucket had been withdrawn in case it required attention when the mine was flooded above the level of the access door. Finally, below the clack piece was the suction pipe or wind-bore, the combined lengths of these two sections being so determined that the suction lift did not exceed 25ft when the pump bucket was at the top of its stroke.

Unlike those fitted to the intermediate pumps, the lower end of the wind-bore in the mine sump was deliberately constricted to produce a powerful suction and so reduce the inequality of load in the eventuality of air being drawn in. It terminated in an elongated bulb pierced with holes and the suction could be varied according to need, either by stopping some of the holes with wooden plugs or by adjusting a sleeve of leather over the bulb. The sound produced by a large beam pump as it drew through these holes was tremendous. To miners who depended for their livelihood on the reliability of the pumps it may have been a familiar and reassuring noise, but to the uninitiated it was awe-inspiring to hear, reverberating through the cavernous darkness of the mine galleries, a sound like the stentorian breathing of some sleeping giant. It was because of this characteristic sound that these holes were aptly named 'snore holes'. In deep mines the lowest pump was generally given a short lift. This was an additional precaution because it minimized the reduction in load which would result if the pump drew air.

The only purpose other than mine pumping to which Newcomen's engine was applied during the inventor's lifetime was that of supplying water to towns, the engines at Passy, near Paris, and York Buildings being the first examples. Here, instead of being lifted from a great depth, the water had to be forced upwards and this necessitated a totally different arrangement of pumps. Two pumps were installed side by side, the one a jack-head lifting pump of the type used to raise water to the injection water cistern and the other a plunger forcing pump as patented by Sir Samuel Morland and first produced commercially in Great Russell Street, London, by Isaac Thompson. The crosshead linking the two pump rods to the arch-head chains was guided by grooves cut into two upright timbers. The jack-head pump rod was of round section iron, turned and polished where it passed through the leather pump gland. The middle portion of the force pump rod consisted of two wooden planks holding between them weights of lead or pig-iron. The engine raised the jack-head pump bucket on its power stroke, while the descending weights actuated the force pump and returned the engine piston to the top of the cylinder. The delivery pipes from the two pumps were led into a closed cistern or receiver in which the compression of air acted as a balancer, forcing the water in a steady stream up a rising main to a storage reservoir placed at a sufficient height to provide enough head for the town supply pipes. According to Triewald the York Buildings engine pumped to a reservoir containing 'several thousand hogsheads' from which water was conducted through lead pipes to every floor of 'over 500 many-storied houses' in the vicinity of Hanover Square, now Hanover Place off Long Acre. On each floor, he tells us, two taps were provided, one for domestic use and the other a 'fire tap' with a threaded outlet to which a leather hose could be attached. Such a precaution reminds us that the great fire of London was then still recent history.

So concerned was Triewald to 'sell' the new power to his countrymen in Sweden that he made claims for Newcomen's engine the extravagance of which would have shocked the inventor. His most gross exaggeration reads as follows:

> As to the durability of this machine it certainly possesses no small advantage over other artifices; because the noblest part of the machine are made of metal, copper, lead or iron and ought thus, as a matter of course, to be able to defy time, nay, a cylinder, after a hundred and even a thousand years' use, is better, and never can be worse.

So far was this from the truth that some engine cylinders required renewal within Newcomen's lifetime. This might not be due solely to bore wear but to casting defects disclosed either by wear or by the 'working' of the cylinder on its wooden bearers when under load. Of the twenty-three iron cylinders supplied by the Coalbrookdale Company between 1722 and 1733, when the patent monopoly ended, by no means all were for new engines. Undoubtedly some were supplied to replace existing brass cylinders. Thus the iron cylinder supplied to Griff in October 1725 was possibly a replacement and it is significant that a brass cylinder was subsequently sold off from Griff. The iron cylinder supplied for the Dudley Castle engine in 1736 was certainly a replacement and the old brass cylinder was later sold for scrap.

This substitution of brass by iron cylinders was roundly condemned by Desaguliers. He wrote:

> Some people make use of cast Iron Cylinders for their Fire-Engines; but I would advise nobody to have them, because tho' there are Workmen that can bore them very smooth, yet none of them can cast less than an Inch thick, and therefore they can neither be heated nor cool'd so soon as others, which will make a Stroke or two in a Minute Difference, whereby an eighth or a tenth less Water will be raised. A Brass Cylinder of the largest Size has been cast under 1/3 of an Inch in Thickness; and at long run the Advantage of heating and cooling quick will recompense the Difference in first Expense: especially when we consider the intrinsick Value of the Brass.

Although this argument is perfectly correct as far as it goes, it could only have been advanced in this way by a scientific theorist totally ignorant of those practical engineering considerations which, from that day to this, have modified ideal solutions. In theory, the thinnest possible brass cylinder wall was undoubtedly the best way of minimizing the thermal inefficiency of the Newcomen working cycle, but if, as Desaguliers says, a brass cylinder was produced with walls only $\frac{1}{3}$in thick, it could never have withstood the mechanical stresses to which it would be subjected. Even if it did, the tolerable wear would be negligible before failure occurred. It seems likely that the brass cylinders commonly used were not less than an inch thick in the wall.

Although only the weights and not the diameters of the iron cylinders cast at Coalbrookdale during the term of the patent are quoted in the company's records, some are known from other sources and it is evident that they did not substantially exceed in diameter the brass cylinders which they superseded or replaced. In these circumstances the argument advanced by Desaguliers carried weight and the iron cylinder made headway solely by virtue of its cheapness in those areas where the Proprietors held sway. As we have seen, in Northumberland and Durham where brass cylinders of better quality and lower price could be obtained from local foundries, iron cylinders were not favoured.

The tide did not finally turn in favour of the iron cylinder until after Newcomen's death and after the patent monopoly had expired. It did so because the progressive deepening of mines led to a demand for much larger and more powerful engines. Brass cylinders of such size would not only have been enormously costly but, in order to withstand the mechanical stresses, their wall thickness would have had to be increased with a consequent loss of thermal advantage over iron. The pioneer Coalbrookdale ironworks was uniquely equipped to meet this demand and for more than forty years appears to have enjoyed a monopoly. The cylinders cast at the Dale increased rapidly in size, until in 1761 they produced a cylinder $74\frac{1}{2}$in diameter and 10ft long for the Warmley Brassworks, near Bristol. Such monster engines were the achievement of Newcomen's successors whose work is the subject of the final chapter.

CHAPTER 6

Newcomens Successor's

In March 1733 Abraham Darby's son-in-law, Richard Ford, who managed the Coalbrookdale works after the death of the former, wrote to his partner, Thomas Goldney: 'As ye patent for ye Fire Engine is about expiring, that business will consequently more increase'. That the Newcomen engine would become a more attractive proposition, once it had been freed from the burden of royalty payments, with a consequent brisk demand for iron cylinders and other cast parts, was a reasonable assumption by Ford, but for a time his expectations were only partially fulfilled. There was certainly a great increase in the demand for iron pipework and other small parts, but over the next fifteen years only just over sixty cylinders were cast at the Dale.

The two areas of England where the services of the Newcomen engine as a mine pump were most needed were the Tyne basin and Cornwall. The Tynesiders clung to their locally-cast brass cylinders while in Cornwall the introduction of the engine was slowed down by the high cost of coal. This may explain why, at first, the demand for iron cylinders was not as brisk as Ford anticipated when the patent lapsed.

We know from subsequent records that during the first twenty years following the lapse of the engine patent more engines must have been built in the Tyne district than in any other part of England. At least seven iron cylinders were supplied by the Coalbrookdale Company in the 1730s to collieries in the North East but there were many small engines with brass cylinders and, because of their limited power, the more waterlogged pits in the area accumulated a multiplicity of engines, some of them sharing the same shaft, in the effort to keep the waters at bay. Thus Jesmond, Heaton and Tynemouth Moor Collieries each had four engines, Long Benton five and Byker no less than six. This may have been good business for engine erectors and for those who produced the parts, but for the colliery owners, it was almost as ruinous as a drowned-out pit.

Larger and more powerful engines were the answer to the situation, and this also meant larger iron cylinders. The man responsible for introducing them to Tyneside was Isaac Thompson, a relation by marriage of Abraham Darby and agent for the Coalbrookdale Company in Newcastle. In 1752 Thompson imported from the Dale a 47in iron cylinder and succeeded in selling it to William Brown for a new engine which the latter proposed building at Throckley Colliery. This was a notable success, for Brown was by far the most celebrated and influential engine builder in the north. Where Brown led, others followed and from this date forward large iron-cylindered engines became the rule, not only on Tyneside but elsewhere in the north. Brown wrote and told his friend Carlisle Spedding in Cumberland about his proposal to use an iron cylinder at Throckley and Spedding was soon asking Brown to obtain an estimate from Thompson for a 42in cylinder for one of his engines at Whitehaven. Spedding already had two engines of this size. He had fitted a 42in cylinder to the first Saltom engine by 1736/7 and the second Saltom engine was built with this size of cylinder in 1739.

Matthias Dunn in his *View of the Coal Trade of the North of England,* (Newcastle, 1844) states that between 1756 and 1776 William Brown had been responsible for the erection of no less than twenty-two engines in the Northumberland and Durham coalfields and that he had built in addition three engines in Scotland, two at Bo'ness and Pittenweem in Fifeshire and a third near Musselburgh. Brown appears to have been amongst the first engine builders to appreciate the need for adequate boiler capacity and to realise that for reliable operation it was necessary to duplicate boilers so that the engine could be kept at work while one boiler was under repair. The engine erected by him at Benwell Colliery in 1762 was supplied by three boilers, while in the following year he was responsible for the 74in engine at Walker Colliery which was fed by four boilers and was then one of the largest engines ever built. The cylinder, which was delivered by sea to Wincomblee coal staithe, weighed 6½ tons and its bore was described as 'perfectly round and well polished'. The engine raised water 89 fathoms in three lifts from a depth of 100 fathoms, the deepest yet reached in the Tyne district. Working at a rate of from eight to ten strokes a minute it consumed 6½ tons of coal a day. It is interesting to note that the boilers used were still made of copper, three having lead domes, but the one directly beneath the cylinder having a copper dome. Three boilers were kept in steam and the fourth spare.

A year later an even larger cylinder of 75in was supplied by Coalbrookdale for Benwell Colliery. Along with parts for three other large engines, a 60in, a 64in and a 66in, it was shipped from Bristol in two lots, one of 35 tons the other 20 tons.

Despite the prominence given to the Walker Colliery engine it was not the largest in existence in 1763. The credit for this goes to an engine from Coalbrookdale, built for the Warmley Brass-works, near Bristol, two years earlier. The cylinder, 74½in diameter and 10ft long, weighed almost 6½ tons and the bottom 2½ tons. The engine worked four pumps, each 29in diameter, to raise water 18ft and return it from the tail of a watermill into a pond so that it could be re-used by the waterwheels. Even at its slowest rate this engine lifted over 9,000 hogsheads (nearly half a million gallons) an hour.

The hey-day of the large atmospheric engines was the 1760s and '70s for as soon as Watt's pumping engines had proved themselves they superseded the simpler Newcomen type for those very deep and wet mines where the maximum power possible was required. Towards the end of the century the atmospheric engine was used mainly for the less onerous pumping and winding tasks. By 1775 a 76in cylinder engine had been built at the Francis Wood Colliery at Wednesbury in South Staffordshire. There is an unconfirmed report of a water-returning engine of fantastic size built in 1789 for Thackeray and Whitehead's cotton mill at Garratt, Manchester, by Bateman and Sherratt, which is said to have had a cylinder 120in diameter, 18ft stroke, working six strokes per minute and raising 3,000 gallons of water per stroke to a height of 13ft. Discounting this engine the largest one known was built in 1810 to pump from the William Pit at Whitehaven; the cylinder was 80in diameter with an 8ft stroke.

In 1769 William Brown, probably at the request of John Smeaton, made a list of engines which to his knowledge had been erected in the North of England and in Scotland. This is most interesting and totals ninety-eight engines of which fifty-seven were said to have been at work when the list was compiled. The cylinder sizes vary from 13in to 75in with no less than twenty-one engines over 60in diameter. Although twelve of the engines listed are not from the North East coalfield, the list does indicate the rapid swing to large engines in that area. Brown does not by any means give the full story as recent research has shown that over 152 engines are recorded as built in the North East alone before that date.

Even if the demand for large iron cylinders had arisen in the days of Newcomen and the engine Proprietors, it is doubtful whether the Coalbrookdale Company could then have satisfied it. Pressure of demand undoubtedly stimulated the development of technique in foundry and boring mill but to some extent the rate of progress was checked by production limitations. It would seem that the earliest Coalbrookdale cylinders were bored on a relatively light boring mill and that cylinder size was governed by the limitations of this machine. The breakage of boring bars gave considerable trouble when larger sizes were attempted and in 1734 a new boring mill was installed at the Dale for which Richard Ford ordered a wrought-iron boring bar 12ft long and 3in in diameter from a Bristol anchor smith. This mill was evidently successful for a second similar bar was ordered for it in 1745. The development of Newcomen's engine in the eighteenth century must be seen against this background of slow but certain improvement in workshop practice.

In 1730 the Cornish mine captains petitioned the Government to remit the duty of 5s 5d a chaldron which was then levied on coal. They maintained that this duty, added to the high cost of importing coal to Cornwall, made the pumping of their mines by engine power economically impossible. This plea was not heeded until 1741, by which time it is said that only three steam engines were still at work in the Duchy. From this date forward however, Newcomen engines in Cornwall began to increase rapidly both in number and size, from three mines with 47in engines in 1746, to a 70in engine at Herland Mine in 1753. Borlase, writing in 1758 remarks upon the number and size of the engines then at work and refers to engines at North Downs mine, Redruth (2), Pit Louran, Redruth (2), Polgooth, Wheal Reeth, Bullen-garden, Dolcoath, the Pool, Bosproual and Wheal Rose. In 1769 when the north country list was compiled, John Smeaton collected particulars of eighteen large engines then at work in Cornwall, eight of them having cylinders exceeding 60in diameter. Nine years later it was stated by Pryce in his *Mineralogia Cornubiensis*, that more than sixty engines had been built in Cornwall since the coal duty was remitted in 1741 (although at least another ten engines built between these dates can be accounted for now) and that many of these had subsequently been rebuilt and enlarged.

The most notable Cornish engine builders at this time were Jonathan Hornblower (son of Joseph, Newcomen's associate), John Nancarrow (who had provided Borlace's list of engines) and John Budge. The engines built in Cornwall were of a higher standard than those in the north of England, for fuel economy was all-important and provided a great incentive for engine builders to strive for greater efficiency. Moreover, Jonathan Hornblower disseminated a wealth of experience which he had inherited from his father.

It is appropriate that it should have been Josiah, a younger brother of Jonathan Hornblower, who was responsible for introducing the Newcomen engine to the New World. This historic engine was ordered in 1748 or 1749 by Colonel John Schuyler who, with his two brothers, owned a copper mine in what is now North Arlington, New Jersey. Copper had been found on the Schuyler estate in 1715 and was profitably worked by driftways until 1735 when it became necessary to sink a shaft. The ore was exported to the Bristol Copper and Brass Works where it fetched £40 a ton. When the shaft reached a depth at which the water could no longer be cleared by horse power, John Schuyler made the inquiries in London which led to his order.

Josiah Hornblower was chosen, perhaps by his better known elder brother, to erect the engine and on 8 May 1753 he set sail from Falmouth on a coasting ship bound for London where the engine parts, many in duplicate and some in triplicate, had been gathered ready for shipment.

With the engine and its erector on board, the American ship *Irene* sailed from London on 6 June 1753 and encountered such rough weather and adverse winds in the North Atlantic that she did not reach New York until 9 September. The rigours and perils of the crossing were such that Hornblower swore he would never make an ocean voyage again. Much to the sorrow of his family in Cornwall he kept his vow and never returned.

At New York the engine was transhipped to a smaller craft which carried it through Newark Bay and up the Passaic River to an unloading point at Belleville opposite the mouth of the Second River. It was then carted overland for about a mile to the head of the mine shaft which was located near the junction of Belleville and Schuyler Avenues in North Arlington. So arrived the first steam engine in the American continent, an event less celebrated than the landing of the Pilgrim Fathers but no less pregnant with significance for the future.

All the engine parts were on site by the end of September 1753, but erection was a slow and laborious job for Hornblower. Stone for the engine house had to be quarried from the mountains and trees felled for the engine beam and other timber work. The accounts show that no less than 211 days were spent in carting stone, timber and clay for bricks to the site, and it was not until March 1755 that the engine was set to work. The engine appears to have worked well and was twice rebuilt following damage by two successive fires before the mine was finally abandoned in the early years of the nineteenth century. It would appear that at each rebuilding a new cylinder was fitted, the first two being of brass and the last of iron. The diameter of the brass cylinder is not stated, but on the 1889 evidence of Mr Justice Bradley of Washington, who married a grand-daughter of Hornblower, the last cylinder 'was of cast-iron, an inch or more in thickness, nearly eight feet long and more than two and one-half feet in diameter'. A relic preserved in the Smithsonian Institution, the United States National Museum at Washington, is believed to be the lower half of this cylinder. The engine is said to have pumped at the rate of 134 gallons per minute from a depth of 100ft using a 10in diameter pump barrel, an iron lifting pipe in 8ft sections and wooden spears.

Josiah Hornblower married and settled in the district where he was associated with the mine until 1794. In that year he built for the last owners of the mine, Messrs Roosevelt, Mark and Schuyler, on land which he sold to them on the outskirts of Belleville, the first ore stamping mill in America. Here, too, the mine owners established a foundry and machine shop where the first steam engine to be manufactured in America was made. This works was named Soho after its famous counterpart in England, but the historical link is not with Boulton and Watt but with Thomas Newcomen through Josiah Hornblower and his father, Joseph. Josiah Hornblower died at Belleville on 21 January 1809 in his eightieth year. According to C. W. Pursell's *Early Stationary Steam Engines in America* the only other atmospheric engines built in America were at a Philadelphia distillery in 1773 (which was never completed), at New York Waterworks about 1774-6 and at an iron mine in Rhode Island about 1776.

Another Cornish engineer who emigrated to America was John Nancarrow, one of the most notable engine erectors of his day. Although still in Cornwall in 1757, by 1786 he was operating a steel furnace in Philadelphia when he was consulted about building an engine for a steamboat.

James Brindley was associated with the Newcomen engine in the mid-eighteenth century. In 1756 he erected a 36in engine for Thomas Broade at Fenton Vivian in Staffordshire. According to details of this engine which Carlisle Spedding sent to his friend William Brown, Brindley mounted his cylinder in the wall of the engine house opposite the lever wall. It would seem that Brindley did this, not because he was aiming at a more rigid structure, but because his patent boiler could not be conveniently accommodated within the engine house. This boiler consisted of a brick vault 18ft long with a floor of cast-iron plates over four small furnaces. Each furnace had its own iron flue pipe which passed back through the water in the boiler before entering the chimney. The parts for the engine were supplied by the Coalbrookdale Company, and their accounts show that Brindley was still experimenting with his boiler in 1759. It was not a success. Although Brindley erected a number of engines in North Staffordshire and elsewhere he was relatively little involved in steam power. Compared to the many other engine builders, such as Wise, Nancarrow, Hornblower, Budge, Curr, Brown, Smeaton, Thompson and other who are even less well-known today, Brindley's contribution was very small indeed, and he is best remembered as a canal engineer.

A somewhat similar experiment was made in Cornwall with the object of utilizing the waste heat from copper ore smelting and roasting furnaces. The idea was patented by the Cornishman Sampson Swaine in July 1762. The boiler was built of 'moorstone' (granite) blocks with a brick vault and according to notes later made by John Smeaton it was 20ft long, 9ft wide inside and 8½ft deep. The hot gases from three reverberatory furnaces were carried throughout its length by three copper flue tubes of 22in diameter and then passed up three chimney stacks. It was built to supply pumping engines at Wheal Kitty and Wheal Chance at the Weeth, Camborne and was judged a failure, principally because the copper flue tubes passed through the water at too high a level. Nevertheless, when the Weeth shafts were abandoned in 1770-1 the boiler was removed and re-erected at Dolcoath mine. Some of the granite blocks which formed the boiler sides are now preserved in the Holman Museum at Camborne.

Smeaton was evidently intrigued with Swaine's idea, for a similar boiler was later built to supply his engine at Chacewater. The boiler shell was made of iron plates instead of stone and brickwork and there were four flue tubes instead of three, but the dimensions were almost identical. The Chacewater engine had three orthodox boilers, one beneath the cylinder and two at the side, and the waste heat boiler was used in lieu of the side boilers.

John Smeaton (1724-92) was unquestionably the greatest of Newcomen's successors. Unlike James Watt, Smeaton did not introduce any revolutionary change in working principle, but he was the first engineer to make a scientific study of

58 The 66in atmospheric engine designed by Smeaton for Kronstadt Docks, which was set to work in 1777.

No. 1 Upright Secti-on for the Fire Engine at Cronstadt
Floor
Floor
7 Lengths of 8.9¼ each
Scale ⅓ of an Inch = 1 Foot
1 2 3 4 5 6 7 8 9 10 11 12 13 14 15 Feet
Level of the intended Ground Floor of the Engine
Level of the Tops of the Stone Pillars of Mill
Pillar

Newcomen's engine. As a result he was able to make a number of detail improvements that together made the machine as perfect as the techniques of the time and its inherent inefficiency allowed.

Before Smeaton's day even the most celebrated engine-wrights such as William Brown or the Hornblowers worked empirically, following the precedents laid down by their predecessors. Practical men, they might know by experience that the engine would work better if such-and-such were done, but they had no idea *why* this should be so and they lacked that capacity for scientific inquiry which would have enabled them to find out. On the other hand a scientist such as Desaguliers could pontificate correctly about the relative inefficiency of iron cylinders and yet, as his writings show, remain completely fogged about the practicalities of design and construction. By contrast, the combination of practical and scientific talent which Smeaton brought to bear upon the Newcomen engine was a sign of the changing times.

The policy followed by Newcomen's immediate successors could be crudely but accurately defined as 'brute force and bloody ignorance'. A larger cylinder was the engine-wright's unvarying answer to a demand for more power, and load per square inch of piston area was his sole measure of the result he achieved. He took no account of working rate and was at no pains to seek out possible sources of mechanical or thermal inefficiency. Consequently, although engines had become much larger, size for size they were no more efficient than the small engines built during Newcomen's lifetime, indeed some of those built by the less experienced engine-wrights were probably markedly inferior.

Samuel Smiles, in his *Lives of Boulton and Watt,* states that Smeaton was first induced to study the subject by the indifferent working of the early engine near his home at Austhorpe near Leeds. However, this engine ceased working before he was born and had certainly disappeared long before he grew to manhood. We know that curiosity prompted him to make inquiries about it and in this way his interest was doubtless quickened. In 1765 he made in his workshop at Austhorpe a model engine for experimental purposes which was unusual in several respects. It was portable, it had an oscillating wheel instead of a beam, the boiler had an internal flue and the cylinder was small in diameter in proportion to its length, being 10in diameter and 36in stroke. The immediate reason for these experiments was the proposal to erect a pumping engine at the New River Head, Islington, to lift water up to a reservoir in order to supply houses at a higher level than the original terminal pond of Sir Hugh Myddelton's New River. The engineer of the New River Company, Robert Mylne, was a personal friend of Smeaton and had consulted him about the project.

The New River Head engine was completed in 1767 but it failed to fulfil Smeaton's expectations. The main reason for this was that Smeaton had been misled by his experiments into the belief that the high stroke/bore ratio, which he used in his model and perpetuated at Islington, would prove more efficient. It was this disappointment which led the thoroughgoing Smeaton to undertake a full-scale investigation into the efficiency of Newcomen engines in the North and in Cornwall. Hence the statistical data compiled by William Brown. No doubt Smeaton was similarly assisted by Hornblower or Nancarrow in Cornwall. In order to assess and compare the performance of these engines Smeaton evolved a standard which he called the duty. This consisted of the amount of water in millions of pounds which could be raised one foot high per bushel (84lb) of coal consumed. He selected fifteen of the Tyneside engines for special study and found that their loading and performance varied widely. Load per square inch of piston area varied from 5·42 to 10·9 and it did not follow that the heavily loaded engine returned the best duty figure. Nor did it follow that the engine with the largest cylinder was the most efficient or even the most powerful. Thus the largest of the fifteen engines, which had a 75in cylinder, returned a duty figure of 4·59m while a 60in engine recorded 5·88m. Smeaton also calculated for each engine the volume of water raised one foot high per minute and called this the 'great product'. This can be translated into horse-power and when this is done in the case of the two engines just mentioned we find that the 75in engine was producing 37·6 horse-power and the 60in 40·8 horse-power.

Such irregularities made it clear to Smeaton

59 The engine house of an engine built at Garrowtree Colliery in South Yorkshire in 1777, converted into a dwelling.

60 Remains of a castellated engine house at Birdwell, near Barnsley. This had an atmospheric engine built in 1813 and drained ironstone mines.

that there must be serious defects in the construction and management of these engines which warranted further investigation on the spot. What he found fully confirmed his suspicions. He discovered that many cylinders and pump barrels were badly bored so that in some places the pistons ran tight while in others water leaked copiously past them. Either the steam admission pipes were too small or the valve gear did not open the steam admission valve fully, with the result that the ascending piston had to draw the steam out of the boiler. Boilers were often inadequate, badly constructed, inefficiently maintained and wrongly fired. Grates were commonly placed too far below the bottom of the boiler. Smeaton also found that many engines were worked on an excessively short stroke, having regard to the length of the cylinder. This meant that much steam was wasted in filling the bottom of the cylinder to no useful purpose, while at the top an unnecessary amount of water accumulated above the piston, cooling the latter excessively and wastefully. The engines subsequently designed by Smeaton were the fruit of the lessons learnt on this survey.

Smeaton was closely associated with the famous Carron Ironworks, the first in Scotland, which went into production in January 1760. He was responsible for designing the plant there, which was considered very advanced, and he now installed at Carron a new cylinder boring mill. This could produce a truly circular bore, though it could not ensure that it was truly parallel because the boring bar was overhung and the travelling carriage used to support its extremity did not provide sufficient rigidity. For this reason it later failed to satisfy James Watt's exacting requirements, but it was an improvement on its predecessors and the Carron works put an end to the Coalbrookdale monopoly. Even a small

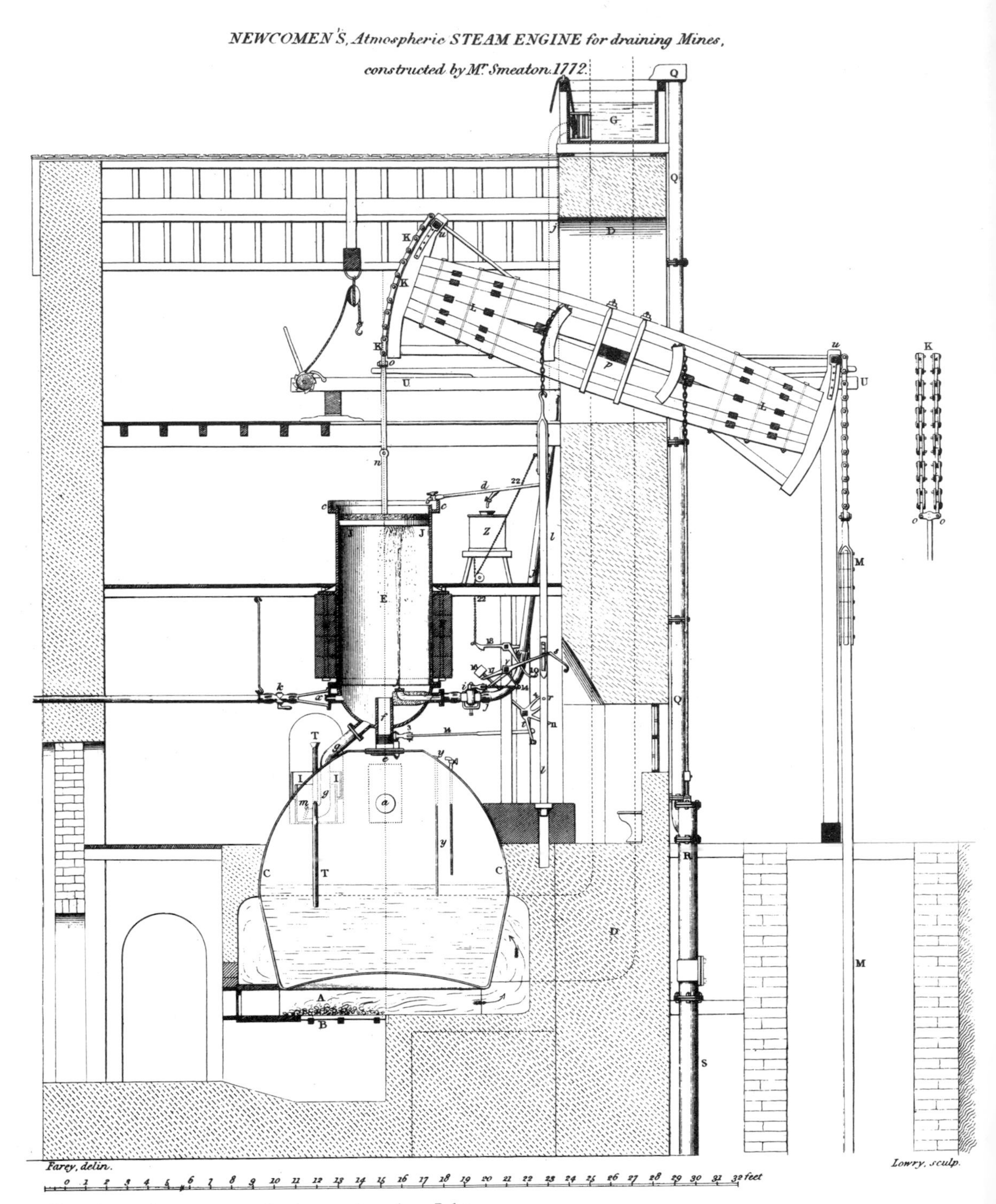

61 Drawing of an atmospheric engine by John Smeaton at Long Benton Colliery, 1772. From J. Farey *A Treatise on the Steam Engine*, 1827.

improvement in cylinder bore accuracy spelled an appreciable gain in efficiency.

Smeaton realized that at the instant of admission to the cylinder the steam should be capable of exerting some force, because it was essential to efficiency that the condensate and air in the cylinder should be expelled through the eduction pipe and the snifting valve as speedily as possible. He therefore ensured that the steam entry was unrestricted and also that the engine was so balanced that the piston offered some initial resistance to the incoming steam. He also appreciated that the admission of injection water to the steam-filled cylinder was equally critical, and that the best results should follow if a finely-atomized spray of water at high velocity could occur the moment the injection cock was opened. To secure this result he decreased the size of the injection nozzle, increased the bore of the supply pipe and raised the cistern to give a greater head. Hitherto it had been the practice to mount the injection water cistern on the beams which supported the spring-beam stages, but Smeaton moved it to the very top of the house, insisting that there should be a fall of at least 36ft from cistern to injection nozzle. He also specified that the injection water pipe, which passed through the cylinder bottom, should be of wood. This was to insulate the cold water in the pipe from the steam in the cylinder. Condensation of steam by the cold pipe was wasteful, while heating of the injection water by the steam was even less desirable. The actual injection nozzle was of brass driven into the end of the wooden pipe. The lower surface of the piston on Smeaton's engines was also insulated with wood to prevent wasteful condensation, the piston being constantly cooled by the water and cold air on its upper side.

Smeaton found that the bad state of the boilers he examined was generally due to the water used being either highly corrosive or so hard that it produced excessive scaling. As explained in the last chapter, it was customary to feed the boilers with injection water from the hot well. Because the amount of injection water used was very great — about ten to twelve times the quantity needed to supply the boilers — it was usually fed to the overhead cistern by a jack-head pump from the mine adit level. It was this mine water which was often so harmful to the boilers. He therefore decided that the boilers of his engines should be fed with rainwater stored in small catchment reservoirs, assuming that no surface supply of suitable quality existed. It was obvious, however, that the great quantity of water used for injection could not be supplied in this way. This was still pumped from the mine, but Smeaton redesigned the hot well tank. It was still placed so that the boilers could be fed by gravity from it, but it now became a true feedwater heater or heat exchanger, containing an inner tank through which the feed-water passed and was heated by the hot condensate.

62 Science Museum Model of Smeaton's 1772 engine at Long Benton Colliery.

Another serious defect of the Newcomen engine which Smeaton appreciated was its lack of flexibility. So long as it was working on a constant load all might be well, but it was extremely difficult for the engineman to maintain such a desirable state of affairs. If the engine began to lower the level of water in the mine, its load would lessen but the engineman had no means of regulating the engine accordingly except by juggling with the pegs in the plug rod which controlled the valve events. This was a hit-or-miss performance, in which much depended on skill, and usually resulted in the engine working at a shorter stroke and very inefficiently. Newcomen had controlled his engines by the buoy pipe arrangement, but now experience had shown that some other form of control was necessary.

Smeaton attacked this problem in two ways. First he introduced a small pet cock on the eduction pipe just below the cylinder bottom through which a limited amount of air could be bled into the cylinder when it was in vacuum. This may seem contradictory when Newcomen was at such pains to rid his cylinder of air, but Smeaton argued that, when an engine was running on light load, it was much more economical to destroy the vacuum partially in this way and so reduce the volume of steam admitted than it was to shorten the stroke and admit the same volume of steam to little purpose. This pet cock was controlled by hand and it was found that the skilled engine-man could regulate its setting to a nicety.

The second innovation which helped to solve this problem was the cataract. This device is said to have originated in Cornwall and it was doubtless adopted by Smeaton as a result of his previous study of Cornish engines. The cataract, or 'Jack in the Box' as the miners called it, was simply a form of water clock. It consisted of a pivoted, weighted lever or tumbling bob enclosed in a small tank. Instead of a solid weight there was a cup or trough into which water flowed from a pipe under the control of a cock. When the cup was full its weight overset the lever and the water then spilled into the tank whereupon the lever returned to its original position. The cock, by regulating the speed with which the water flowed into the cup, set the frequency of the cycle. The second arm of the tumbling bob could be connected by a chain or cord to the detent which released the weighted 'F' lever and so opened the injection cock. It thus controlled the rate of working in the same way as Newcomen's buoy gear. When the engine was working full power the cataract was disconnected but, when maximum output was not required, the engine-man could link up the cataract and regulate the number of strokes per minute according to need by adjusting the flow to its cup. In a more refined form the cataract gear was fitted to single-acting Watt engines in Cornwall, and to the Cornish engines which succeeded them, to control the injection to the condenser.

Smeaton adopted a hemispherical instead of a flat cylinder bottom and introduced his own design of beam. This was built up from a number of longitudinal timbers braced together and was of great depth, the trunnions or 'gudgeons' being placed at the mid-point instead of below the beam as heretofore. This innovation was not successful. Although very strong when new, the timbers tended to 'work' and later builders reverted to the simple beam, sometimes trussed from central kingposts.

The first engine in which Smeaton embodied these ideas (apart from the hemispherical cylinder bottom) was at Long Benton Colliery, Northumberland in 1772. This had a 52in cylinder and worked 7ft stroke. In his first engine at the New River Head he had made the same mistake as Newcomen did initially by loading the piston too heavily in the belief that this would prove the most economical. Accordingly the New River engine was first started with a load of no less than $10\frac{1}{4}$ lb/in² of piston area which was higher than that first proposed by Newcomen. Smeaton soon realized his mistake as others had done before him and his Long Benton engine worked $12\frac{1}{2}$ strokes a minute on a mean effective pressure of $7\frac{1}{2}$ lb/in². This he considered the most efficient compromise and he adopted it on all the many engines he built. The engine returned a duty figure of 9·45m whereas the best figure recorded by any of the Tyneside engines he had previously studied was 7·44m. Smeaton considered this his standard of performance.

Smeaton followed the Long Benton engine with a 72in machine at Chacewater in 1775, which became famous in Cornwall, and by a 66in engine for emptying the docks at Kronstadt, Russia. The parts for the latter, including the three cast-iron tun boilers, were made by the Carron Ironworks.

It should not be supposed that Smeaton's

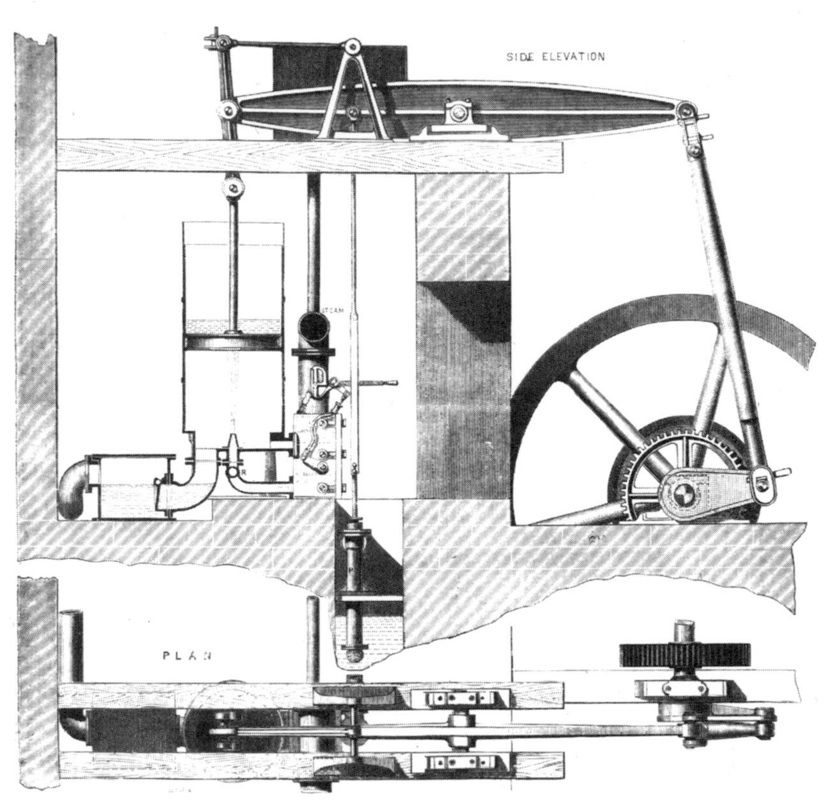

66 Drawing of a Newcomen winding engine at the Farme Colliery, Rutherglen, Scotland, built in 1820. *The Engineer* 1879.

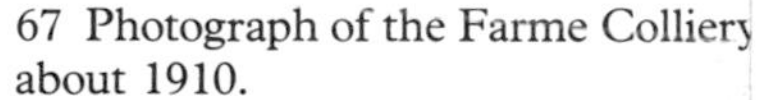

67 Photograph of the Farme Colliery about 1910.

improvements were either quickly or widely adopted. Many engines continued to be built, or re-erected in new engine-houses, by engine-wrights who were either unable or unwilling to learn. It may be recalled that as late as 1811 George Stephenson first distinguished himself as an engineer by setting to rights a Newcomen-type engine which had just been erected at Killingworth High Pit and would not work satisfactorily. Stephenson's answer was to raise the level of the injection water cistern to provide a better head and to enlarge the bore of the injection cock, remedies which Smeaton had first applied forty years before.

Smeaton built many engines for pumping purposes in mines and waterworks, but he was extremely conservative when proposals for the wider application of Newcomen's engine were canvassed. It was natural that as the engine grew in popularity suggestions that it be made rotative should be advanced, but Smeaton insisted that the

63 An early photograph showing a Newcomen-type winding engine in use at a Coalbrookdale colliery about 1860.

64 A Newcomen winding engine at Nibland Colliery, Nottinghamshire, about 1855. Note the lack of arch-head at the piston-end of the beam and the heavy connecting rod at the crank end.

65 A Heslop engine at a colliery in the Madeley Wood area of Shropshire, used both for pumping and winding in 1879. The pump rods were attached to the left-hand end of the beam and a crank rod was attached to the right-hand end working a rope drum.

only effective way of doing this was to make the engine pump water back from the tail-race to the head-race of a waterwheel. Many of these 'water returning engines', as they were called, were built before Watt's rotative engine opened a new chapter in the history of steam power. In the context of his age Smeaton was right because, curious though it may seem to us, even the most eminent engineers were convinced that the non-rotative engine could not be made rotative by the simple addition of connecting rod and crank because its stroke was variable. The fact that a

crank would be capable
was not appreciated; on
believed that the variabl
form of positive connec
rotative motion were
complex and trouble-
toothed racks, ratchet-
wheels. Compared wit
was right in favouring
waterwheel.

Perhaps the first at
motion from a Newco
Triewald in Sweden i
suggested that he may
similar earlier devic
Richardson who wa
Whitehaven visited Lor
an attempt to obtain a
the engine. His meth

advertised his machine in a broadsheet and claimed by raising 660 corves a day, a saving of £250 per year 'besides saving ye trouble and risque of keeping 20 Horses for the same work, which in Countrys where Hay and Straw are scarce often puts the coal owners to great inconvenience.'

Another attempt to make a Newcomen engine rotative was made by Joseph Oxley at Hartley colliery, Seaton Delaval, Northumberland, in 1763. By means of his patent ratchet device he made the engine turn a 'whim' for raising coal from the mine, but its action was so irregular that it was soon converted to waterwheel drive, the engine raising the water to the wheel.

John Stewart patented another ratchet arrangement in 1766 and in 1768 he is believed to have constructed in Jamaica a sugar mill driven by a Newcomen engine, but the gear was complicated and cumbersome and the venture came to nothing. The latest of the 'ratchet and click' patents, as they were called, was that taken out by Matthew Wasbrough, the Bristol engineer, in 1779. Wasbrough supplied one of his patent rotative engines to James Pickard, a Birmingham button maker, and it was Pickard who shortly afterwards took the decisive step of substituting a connecting rod and crank on this engine. Contrary to expectation the arrangement worked, albeit clumsily, and one of the most remarkable fallacies in engineering history was exploded.

Wasbrough and Pickard proposed carrying their engine over centres by the momentum of a weighted wheel on a countershaft geared up two to one to the crankshaft, but this arrangement was rejected in favour of a heavy flywheel mounted on the crankshaft itself. Even so, the orthodox Newcomen engine in rotative form was irregular in action and was therefore not ideal for driving machines which required an even torque, such as those used in the textile industry. Because the piston was still linked by chain to the beam there could be no reversal of load and for this reason, in addition to the flywheel, a heavy counterweight had to be applied to the connecting-rod end of the beam.

In the attempt to overcome these defects a number of double-cylinder rotative engines of Newcomen type were built in the last decade of the eighteenth century. Bateman and Sherrat of Salford built for Lancashire mills a number of engines in which two Newcomen cylinders drove an overhead rocking shaft by racks on the piston rods which engaged a large gearwheel on the shaft. Rotary motion was derived from this rocking shaft by cranks and connecting rod. The so-called 'pendulum steam engine' built by Jabez Hornblower and Maberley in London in 1795 was very similar except that chain connections to an overhead pulley were used instead of the racks and gearwheel. Both these builders, however, soon adopted the separate condenser and so became infringers of the Watt patent.

Undoubtedly the most workmanlike rotative engine was that designed and built by the celebrated Derbyshire enginewright Francis Thompson (1747-1809). Thompson placed his two cylinders one above the other, with enough space between them to admit air to act freely on both pistons, and used a common piston rod. Because the upper cylinder was inverted, the piston rod worked through a gland in what would normally be the cylinder bottom. Triple chains connected the piston rod to the arch-head of the beam; two were connected normally while the third was yoked to the top of a vertical bar which, in effect, extended the piston rod above the arch-head. This third chain pulled the beam up when the upper piston made its power stroke. The other end of the beam was coupled by a connecting rod to the crankshaft and flywheel. Thompson, who patented this engine in 1792, built several of them to drive textile mills in the North Midlands and Manchester. The largest of these, with two 40in cylinders and 6ft stroke, was erected at Davison and Hawksley's mill at Arnold, near Nottingham. It was customary to work rotative Newcomen engines on a light load of from $4\frac{1}{2}$ to 5 lb/in^2 of piston area and at a much faster speed. Thus the Arnold engine made eighteen strokes a minute, turning the 18ft flywheel at $45\frac{1}{2}$ revolutions per minute.

Another type of double cylinder engine which was made in varying forms by Jonathan Hornblower, Sadler, Heslop and Thompson during the closing years of the eighteenth century was really a rudimentary form of compound. Steam was transferred from a closed cylinder to a second Newcomen-type cylinder where it was condensed by the usual water injection. This

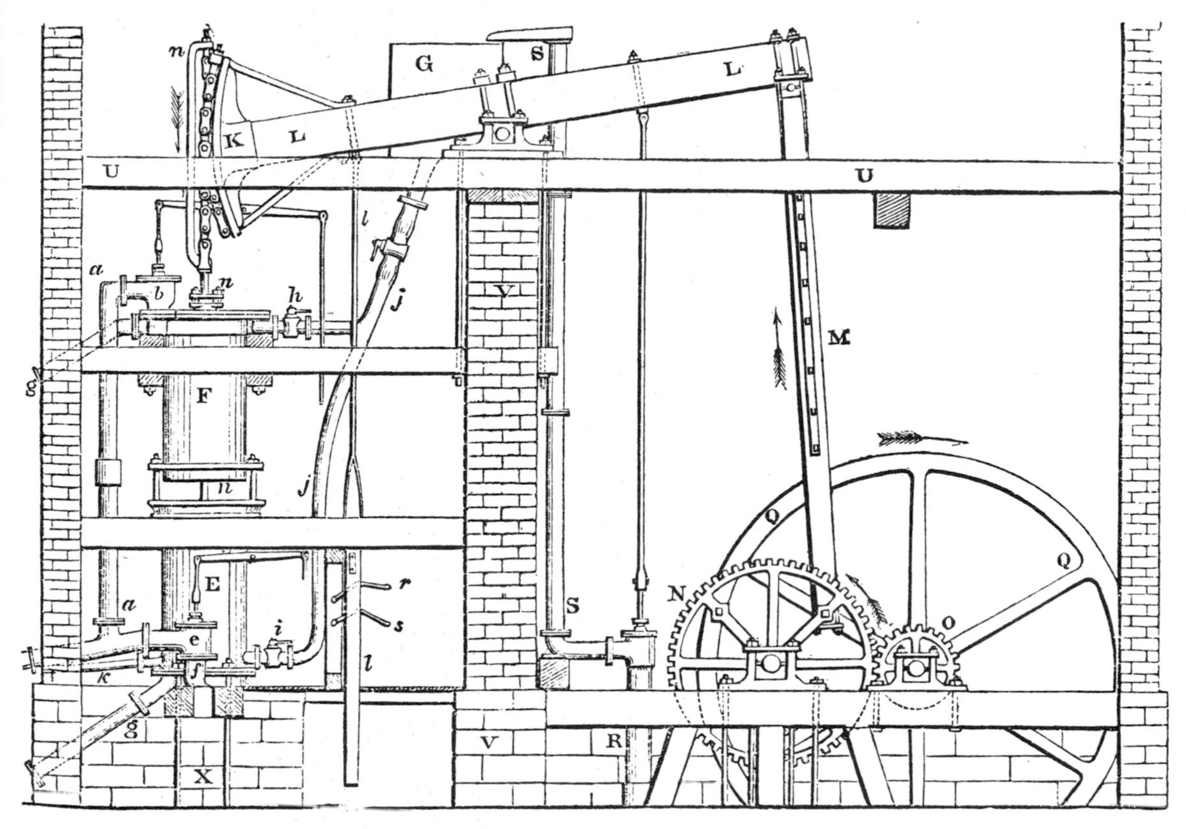

68 Francis Thompson's twin cylinder double-acting rotative engine.

69 A rotative atmospheric engine at Basset Pit, Denby, Derbyshire, in 1885. Another photograph taken earlier shows that the left-hand beam was a power take-off from the flywheel, working pump rods in the pit shaft off the picture to the left.

70 Typical atmospheric or whimsey winding engine as used on the Staffordshire Collieries. These survived in many cases well towards the end of the nineteenth century.

scarcely comes within the terms of reference of this book, however, because in this context the Newcomen cylinder acted as a condenser and air pump to its fellow and engines of this type were therefore held to infringe the Watt patent. Infringement or no, they were certainly evolved as a result of Watt's invention. Small colliery winding engines of Heslop's design remained in use at Madeley, Shropshire and in Cumberland until the 1880s, while the Coalbrookdale Company built and used Sadler's winding engines in the 1790s. Both men worked at the Dale for a time.

An increasing demand for engines came from the growing iron industry. The ironmasters relied upon water power to drive the bellows which produced the blast for their furnaces and, as more furnaces came into blast to cope with a growing demand for iron, this source of power became quite inadequate. First, horse-driven pumps and then Newcomen engines were introduced to return water to the reservoirs and so eliminate shut-downs in dry seasons. The first of these 'water-returning engines' was installed at Coalbrookdale in 1742 and a 72in returning engine was set to work in 1765 at the Carron Ironworks. At first the waterwheels drove large bellows but these were succeeded by cylinder and piston-type blowing engines. Smeaton rebuilt the Carron returning engine, which had proved very inefficient, and at the same time introduced his own design of blowing engine, which had three cylinders driven by a three-throw crankshaft from the waterwheel. At the time of its introduction, this machine was considered the last word in blowing machinery and, for the same reason that Smeaton frowned upon rotative Newcomen engines, he resisted suggestions that the engine might blow the furnace directly and so dispense with the need for reservoirs and waterwheels. Despite Smeaton's doubts, however, the direct, non-rotative blowing engine was introduced about 1784. In place of the pump rods a large blowing cylinder with a weighted piston was connected by chains to the beam arch-head, and a 'wind trunk' led from this cylinder to the blast furnace nozzle via a regulating cylinder with a floating piston to equalize the blast. Later, the blowing cylinder was inverted so that it blew on the power stroke of the engine instead of by the descent of the weighted piston, while a water displacement regulator superseded the piston type. By this time, however, most iron-masters were using Watt engines to blow their furnaces. The Coalbrookdale Ironworks, for example, introduced a Watt blowing engine as soon as they decided to give up the use of waterwheels for this purpose.

When John Smeaton inspected one of the first small Watt engines at work at a distillery at Stratford-le-Bow in April 1777, he described it as a pretty engine but too complicated. This was an honest opinion and not sour grapes. Smeaton was a good friend to Watt and it was on his recommendation that the first Watt double-acting rotative engine to be built commercially was installed at Messrs Cotes and Jarratt's oil cake mill at Hull in 1784. The mill had previously been worked by a waterwheel installed by Smeaton and driven from a reservoir fed from the town's water supply. This supply was furnished by a Newcomen-type engine, which was unusual in having no beam, the pump being connected directly to the piston rod by crossheads and side rods passing on either side of the cylinder. This engine pumped water into the mill reservoir at night when domestic supply was not required, one of the first instances of power storage in 'off-peak' periods. Nevertheless, in the seven years that had passed since he had first seen the Watt engine at work, the broad-minded Smeaton had learned to appreciate its merits and did not hesitate to recommend the engine to supersede his own waterwheel arrangement.

Over a period of more than sixty years men had grown so well accustomed to the Newcomen engine that Smeaton was not the only engineer to shake his head on first sight of James Watt's invention. Watt's engine *was* more complicated and its construction called for much higher standards of workmanship in the making of its valves and valve gear and in the boring of its cylinder. The very fact that the Newcomen engine had been conceived at a time when technology was in its infancy meant that the crudest local resources could build it and make it work, albeit very indifferently. With the new Watt engine it was quite otherwise. The valve assemblies must be supplied by Boulton and Watt, the cylinder bored by John Wilkinson on his new machine at Bersham Ironworks and erection of the engine superintended by a skilled man trained at the Soho works. In other words, engine building began to move out of the hands of the millwright and into those of the specialist manufacturer and the transition was completed when the Watt rotative engine for Cotes and Jarratt was assembled complete and tested at Soho before being despatched to Hull, although this was the exception rather than the rule, even for Watt engines.

The tradition of self-sufficiency and the mistrust of new fangled complexities died hard, especially while Boulton and Watt enjoyed a patent monopoly and demanded substantial royalty payments from users of their engine. Yet the advantages of the Watt engine with its separate condenser which obviated the alternate heating and cooling of the cylinder were so great that it was bound to prevail.

The economy thereby achieved ensured that in a few years Watt swept the Newcomen engine out of Cornwall and even Smeaton's celebrated Chacewater engine was converted to the Watt cycle.

Watt's double-acting engine, introduced about seven years after his single-acting pumping engine in 1776 was admirably adapted for rotative motion. The rotative Newcomens which have been described were no more than ingenious but clumsy efforts to circumvent the Watt patent monopoly. They were almost as complex, less reliable and far less efficient. It was only in the colliery districts of England, where boilers could be fired on unsaleable small coal and economy was therefore of little account, that Newcomen's engine survived stubbornly, not only throughout

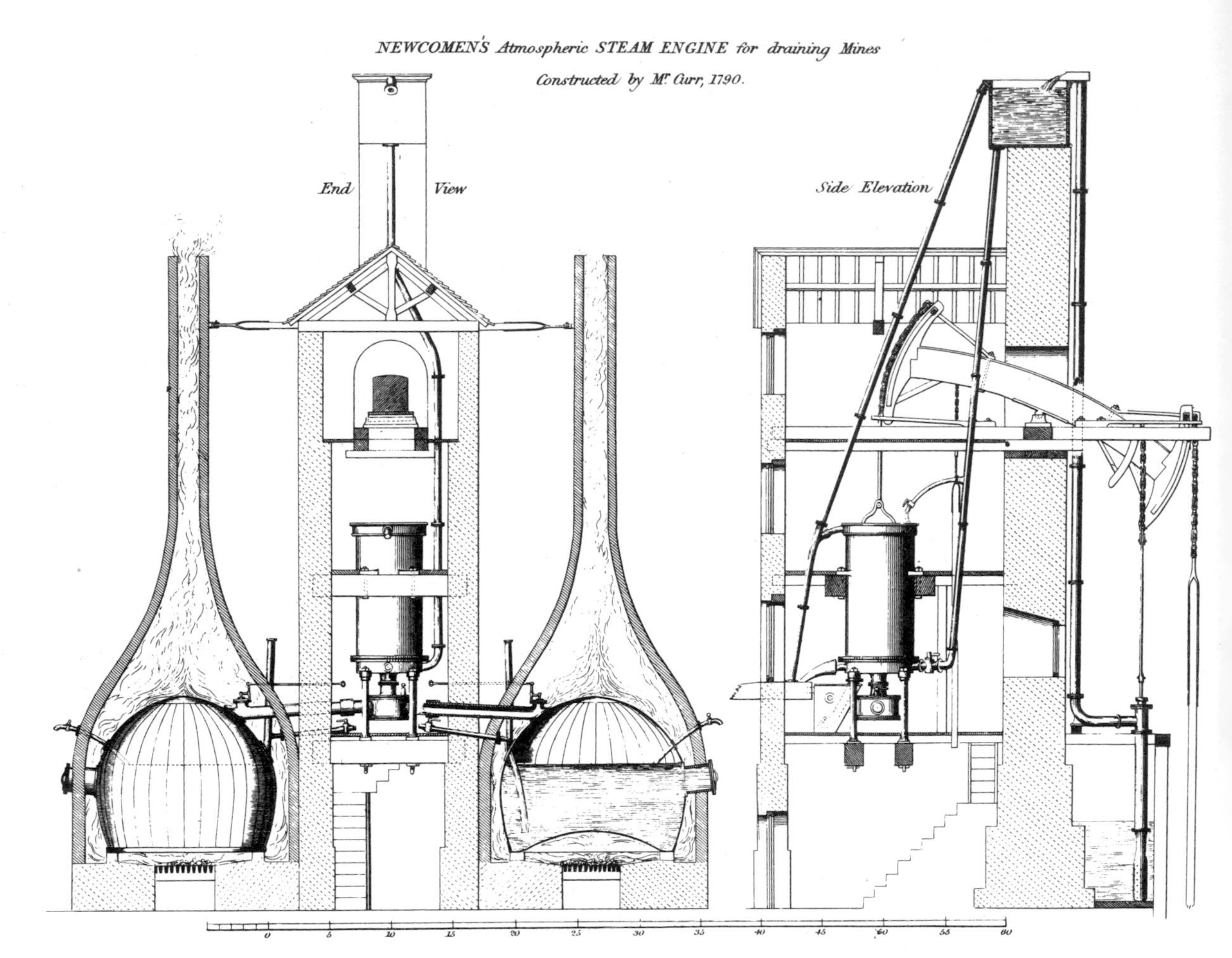

71 An engine with two boilers as illustrated by J. Curr, 1797, *The Coal Viewer & Engine Builders Practical Companion.*

the term of the Watt patent but for long afterwards. The advantages of simplicity and familiarity prevailed when economy could be ignored.

These virtues ensured its use for many years. Lists compiled recently include nearly 1,500 Newcomen engines built before the end of the nineteenth century, while up to the same time there were only about 500 Watt engines.

The most celebrated builders of Newcomen engines at the end of the eighteenth century were the aforementioned Francis Thompson and John Curr of Sheffield. Curr built a number of pumping engines in the South Yorkshire coalfield and published a book on the subject in 1797 called *The Coal-Viewer and Engine-Builder's Practical Companion* which included much tabulated data. In the 61in engine which he built at Attercliffe Common Colliery in 1790, Curr introduced an improvement which was long overdue by banishing the boiler from the engine house. Hitherto, with the exception of James Brindley's engine, even when duplicate boilers had been introduced, it had always been the custom to mount one boiler directly beneath the cylinder just as Newcomen had done. This had always been a structural weakness and a source of trouble which became greater as engines increased in size. The dimensions of the engine house were governed by the diameter of the boiler and its surrounding furnace and flues and this necessitated the use of beams of considerable span to support the weight of the cylinder. The larger the boiler, the greater the span and this may be

72 A Science Museum model of a haystack-type boiler as used on later engines.

73 An old haystack boiler found at Park Hall Farm, near Derby, and once used at Basset Pit, Denby.

one explanation for the tendency to build boilers too high in proportion to their diameter. Inevitably cylinders tended to 'work' on their supporting beams, causing the failure of steam joints, breakage of pipes and generally increased wear and tear. Curr placed the two boilers of his Attercliffe engine in lean-to buildings adjoining the engine house and connected their main steam pipes to a steam chest containing the admission valve directly beneath the cylinder. This arrangement enabled him to contract the walls of the engine house and support the cylinder firmly by two sets of beams, each of very short span. One set extended beneath an intermediate cylinder flange in the usual way, while the lower set carried the bottom of the cylinder on four short columns. It is surprising that an engineer so gifted as Smeaton did not introduce such a desirable modification at an earlier date. Curr also still further increased the height of the injection water cistern, placing it on a pier of masonry extending well above the roof ridge of the engine house. His engine was loaded to 7lb/in² of piston area, made twelve strokes (8½ft) per minute and returned a duty figure of 9·38m.

A similar arrangement of side boiler was used by Francis Thompson in the special circumstances of his remarkable underground engine installation at the Yatestoop Lead Mine in Derbyshire. Thompson built a 70in engine at this mine in 1777, but its adit or sough level was so far beneath the head of the shaft that this engine had to work no less than 540ft of 'dry spears'. In order to obviate this great weight of rods, an underground chamber was excavated about 100ft above adit level over 400ft beneath the surface and within this Thompson erected, in 1782, a second engine with a 64½in cylinder. By excavating this chamber in the softer shale strata at a higher level the cost was much less than if the engine had been erected at adit level, where the strata was hard limestone. Even so, the cost of excavation alone was over £300 before the engine was built. Under these unusual circumstances he decided to supply this engine with steam from a single side-mounted boiler of the haystack type 20ft in diameter. This is believed to have been the largest boiler of its type ever built. It supplied steam to the side of the cylinder through a large bore plug-cock instead of the usual sector plate valve as the

74 Engine drawing on a jug, probably made for Stephen Thompson, brother of Francis.

75 Newcomen-type engine with a 'pickle-pot' condenser built near Stourbridge, Worcs. A drawing made from memory about 1900 by William Patrick, engineman.

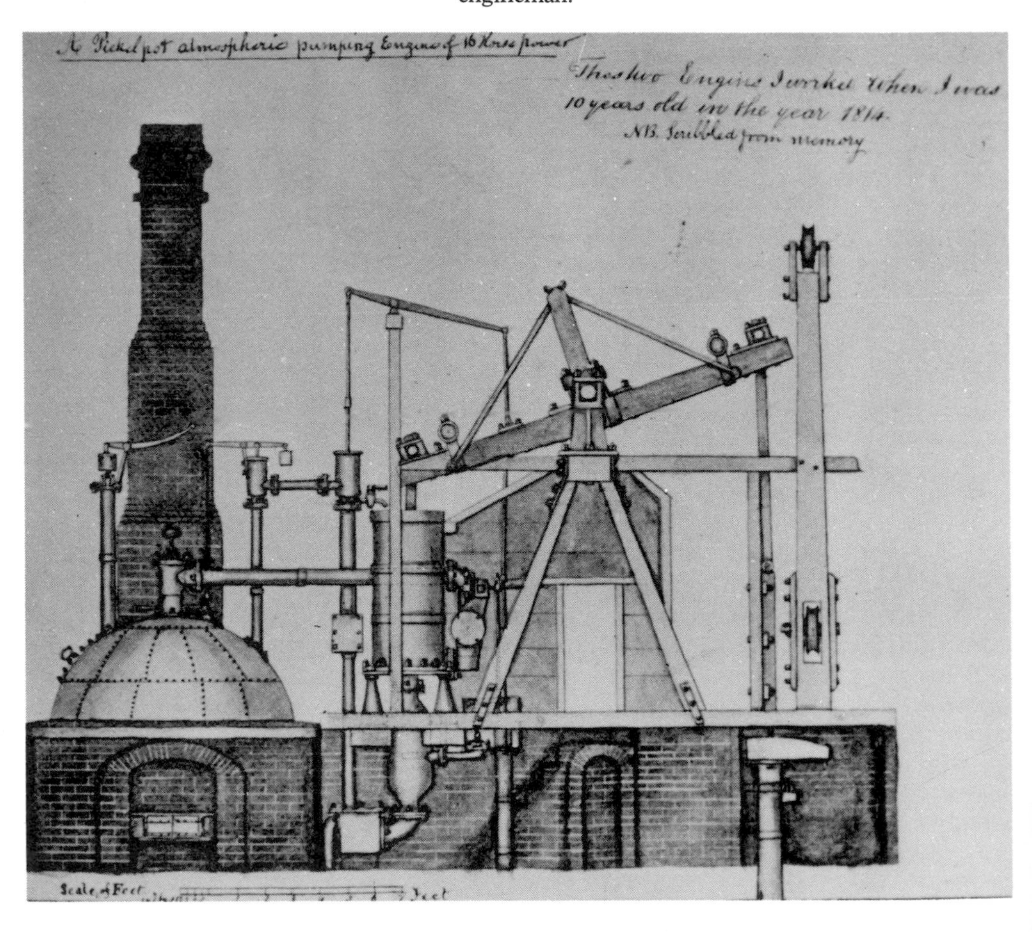

cylinder rested upon a bed of solid masonry. The heat and the noise in this underground cavern when the engine was at work can scarcely be conceived and the ingenuity of man can seldom have contrived anything so closely approximating mythical notions of the infernal regions. Another underground engine was at Whitehaven. This worked inclined pump rods which followed the dip of the coal seams under the sea.

John Curr's use of side boilers was widely adopted during the last years of the eighteenth century and after. Many existing Newcomen engines were converted in this way when they were moved to new sites or when their boilers became due for renewal. Such conversion was often accompanied by the installation of a crude form of separate condenser which became known as the 'pickle-pot'. This was an attempt to evade the Watt patent, although Boulton and Watt held that it was an infringement. It consisted of a pot-shaped vessel mounted below the cylinder and in constant communication with it by a pipe of large diameter. Cold water was injected into the 'pickle-pot' instead of directly into the cylinder. There was no air pump. The incoming steam blew the condensate and air out of the 'pickle-pot' through a non-return valve which was lightly loaded to open at a pressure of approximately 2 lb/in^2. Compared with the Watt condenser this arrangement was very inefficient. Its limited advantage was that when steam was admitted to the engine cylinder it was not immediately exposed to so large a surface area of condensate, the diameter of the 'pickle-pot' being much smaller than the diameter of the cylinder. After the lapse of the Watt patent in 1800, some Newcomen-type engines were fitted with separate condensers of the true Watt type, but others retained their 'pickle-pots' throughout a working life which in some cases, lasted into the present century.

Many of the improved features introduced by Boulton and Watt were incorporated in the design of the atmospheric engines built or re-built after 1800. Watt-type drop valves with their attendant gear replaced the old sector plate valves and plug cocks with their 'Y' and 'F' lever working gear, while cast-iron beams appeared in place of wood. In most cases, however, engines exhibiting these latter-day features prove to be older in origin. When, as frequently happened, the exhaustion of mine workings brought about the removal of an engine to a new site the reconstruction afforded an opportunity to introduce such modifications. Because these removals were rarely recorded, it is extremely difficult to date with certainty the few Newcomen-type engines which survived down to modern times.

76 The Elsecar engine house and pump shaft gear. The pump rods have been cut away and modern weights added to balance the beam.

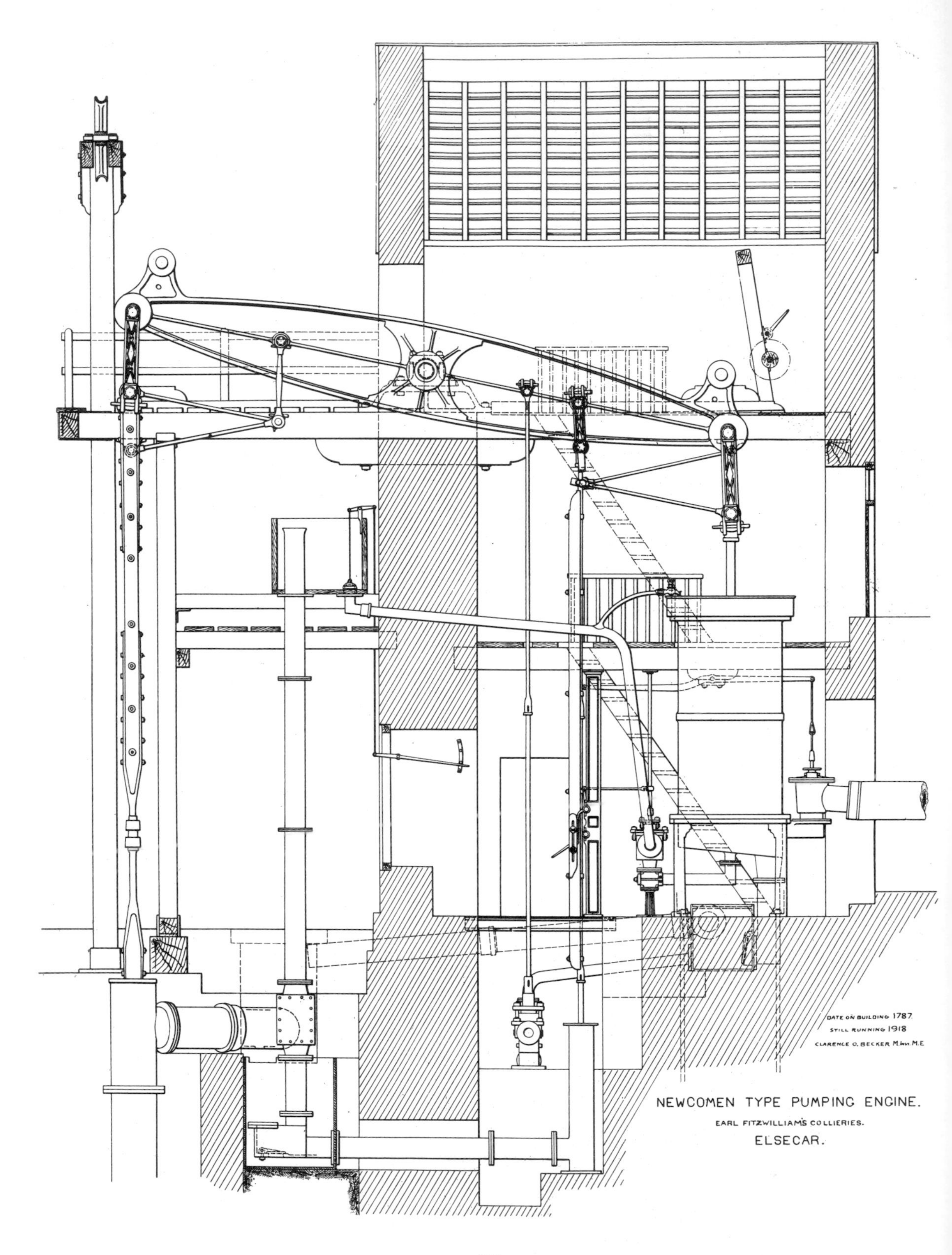
DATE ON BUILDING 1787.
STILL RUNNING 1918
CLARENCE O. BECKER M.Inst.M.E
NEWCOMEN TYPE PUMPING ENGINE.
EARL FITZWILLIAM'S COLLIERIES.
ELSECAR.

77 Drawing of the atmospheric engine at Elsecar, near Barnsley. The engine was fitted with drop valves, parallel motion and cast iron beams in the nineteenth century, but dates from 1795. Superceded by electric pumps in 1923 and now preserved on site.

78 The valve gear of the Elsecar Engine.

79 Detail of the beam of the Elsecar Engine.

The engine built at Elsecar Colliery near Barnsley was erected in 1795 with a 42in cylinder. The date of 1787 on the engine house is incorrect. A new cylinder of 48in diameter was fitted in 1801. The wooden beam was replaced by one of cast iron in 1836 and at the same time parallel motion was fitted both on the cylinder and pump ends of the beam. This was unusual for mine engines but common on waterworks engines. In this form the engine worked continuously until 1923 and in emergencies until the 1930s. Fortunately the engine has been preserved and can be viewed today.

The engine at Westfield, Parkgate, Yorks, like that at Elsecar pumped from the Barnsley Seam workings but was not built until 1823, abundant slack coal being available so that the efficient Watt-type was thought unnecessary. The cylinder was 54in diameter and 10ft in length but it worked with an average stroke of 5ft 9in instead of its maximum of 7½ft. The main beam was of cast iron in two flitches 25ft long by 5½ft deep at the centre and weighed about 8 tons. The valve gear was operated by a plug rod and similar in principle to the earliest engines. The engine worked about 10 strokes a minute from a depth of 60yd. It was dismantled in 1934 after over 100 years of service.

A 36in pumping engine built by the Carron Company about 1775 was moved to Caprington Colliery, Ayrshire, in 1806. About 1837 a cast-iron beam and Watt parallel motion were substituted and in this form the engine continued to work until July 1901.

81 Another very old and much altered pumping engine at Staveley, near Chesterfield, Derbyshire. Originally built in 1776 by the Coalbrookdale Company for the Old Handley Wood Pit in Shropshire it was moved to Staveley in 1849, where it worked until about 1880. A 'pickle-pot' condenser and 'modern' valve gear was added and the original wooden beam replaced by one of cast iron. Note the repair to the cracked beam, and the cataract near the cylinder top for regulating the speed of the engine. (*The Engineer,* 1880).

80 The 36in diameter engine at Caprington Colliery, Stirlingshire, Scotland built about 1775 by the Carron Company had a cast-iron beam and parallel motion fitted later and then worked until 1901.

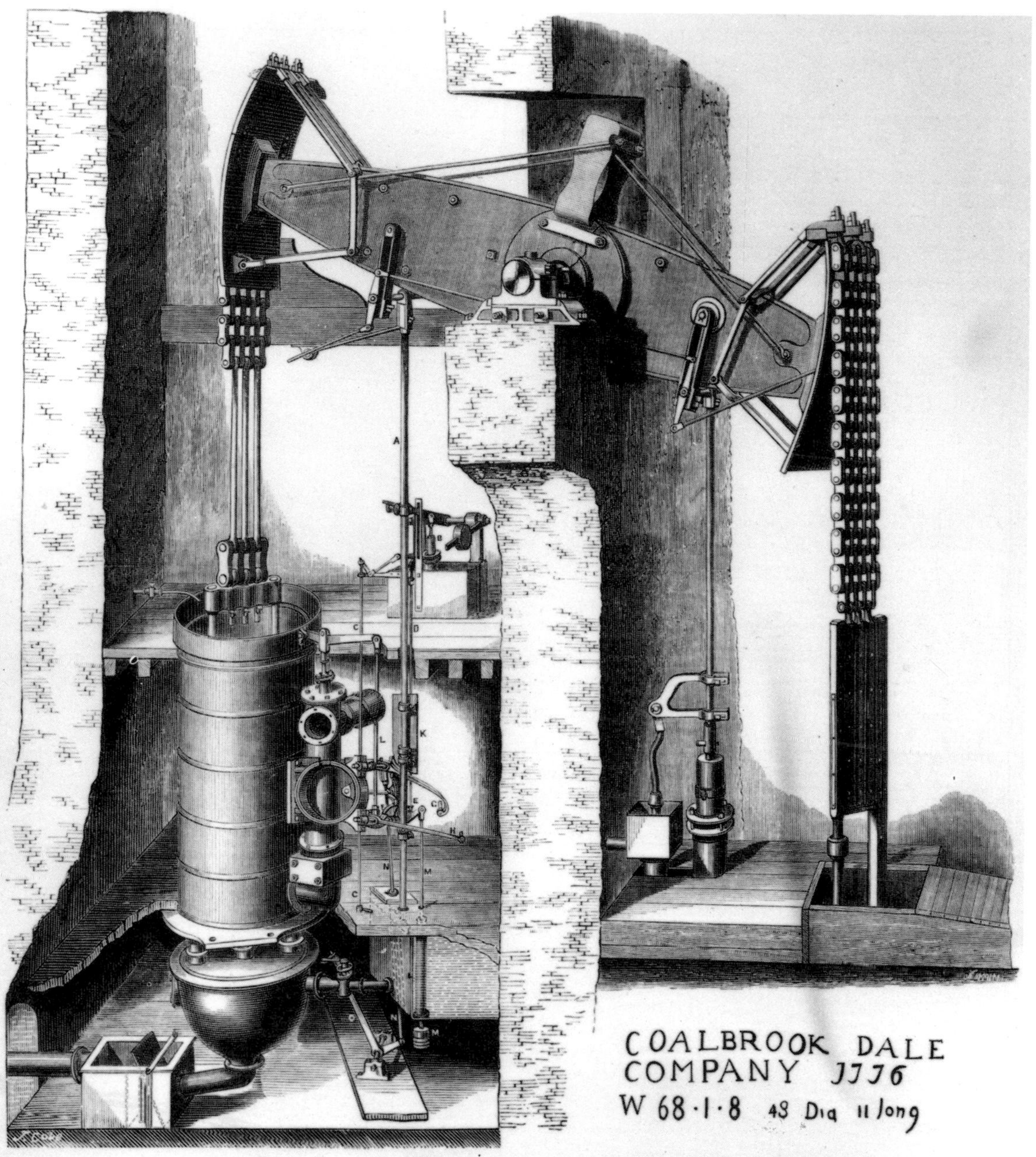
COALBROOK DALE
COMPANY 1776
W 68·1·8 48 Dia 11 long

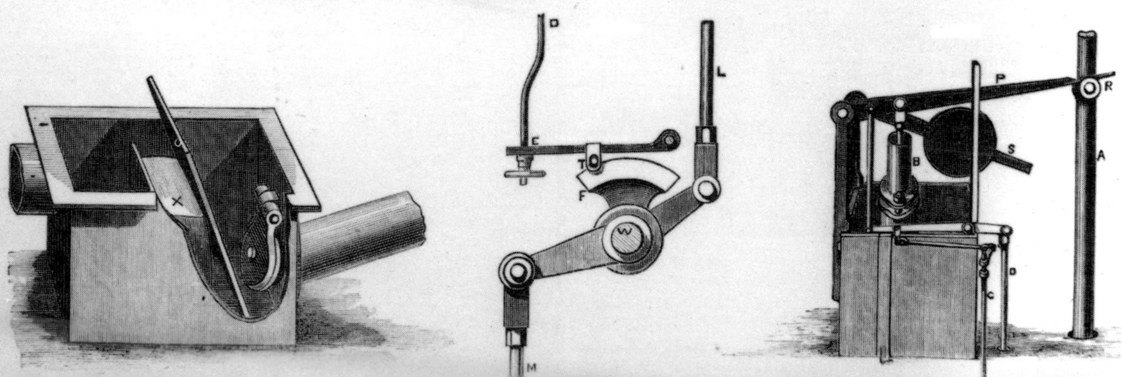

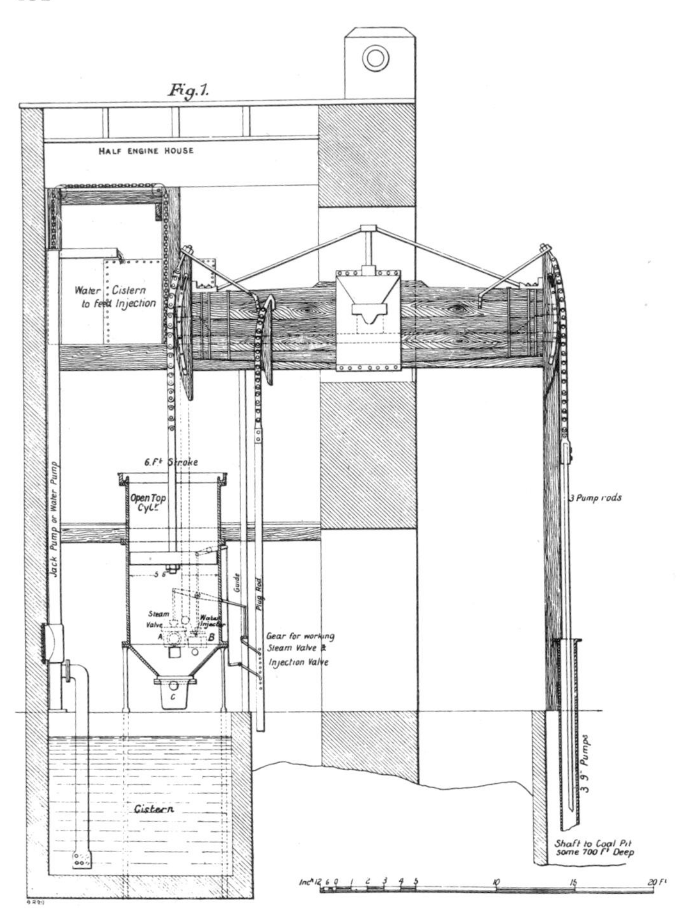

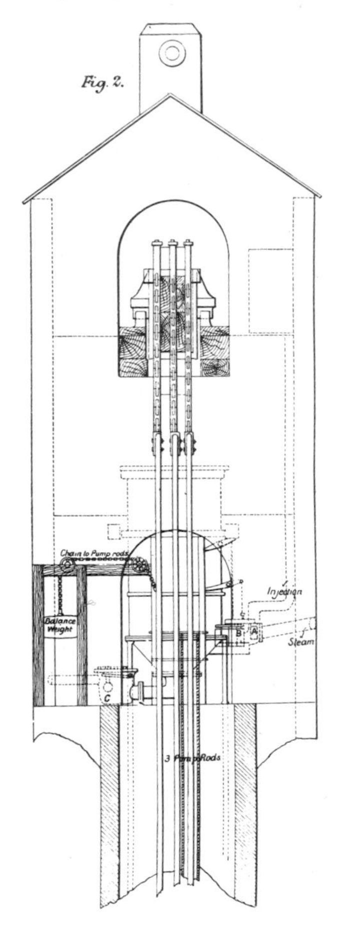

82 Drawing of the old engine at South Liberty Colliery, Ashton Vale Iron Company, Bristol. Worked from mid-eighteenth century until 1900 when it was dismantled. Although the valve gear had been updated this engine was probably unique in retaining its wooden beam into the twentieth century. (*Engineering* 1895).

83 Indicator diagram taken in May 1895 from the Newcomen engine at Ashton Vale, Bristol.

Dia. of Cyl. 5' 6"
Stroke 6' 0" about
No. of Strokes per min. 10.

Boiler Pressure 2·3 lbs.
Vacuum Gauge, none fixed.
Time 3 p.m.

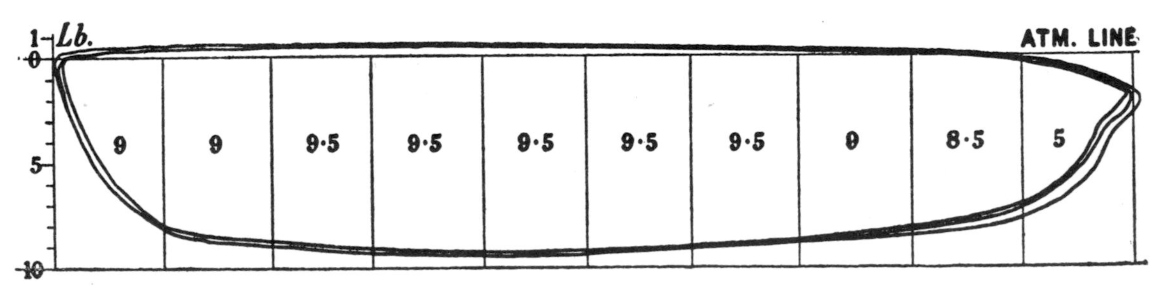

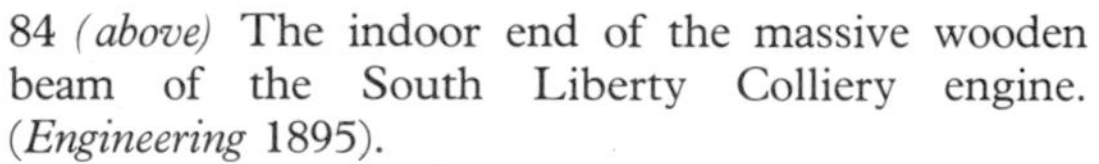

84 *(above)* The indoor end of the massive wooden beam of the South Liberty Colliery engine. (*Engineering* 1895).

85 *(top right)* Another view of the indoor end of the beam. (*Engineering* 1895).

86 *(bottom right)* The outdoor end of the beam with the arch-head and chains connected to the pump rods. (*Engineering* 1895).

The Newcomen-type engine which probably holds the record for longevity in a near original state was the 66in pumping engine at the South Liberty colliery of the Ashton Vale Iron Company, near Bristol, which worked regularly until it was dismantled in 1900. H. W. Pearson, a member of the Institution of Mechanical Engineers, visited the engine in 1895 to record it by means of photographs and drawings. He also took an indicator diagram. The latter recorded a mean effective pressure of between 9 and 10 lb/in^2, showing that at its normal working speed of ten strokes a minute the old engine was giving about 52 horse-power. A drop valve controlled the steam admission and a form of sliding valve was used for the injection, but there was no form of separate condenser and a wooden beam with arch-heads and chain connections to piston, pump spears and plug rod was retained to the last. The engine pumped from a depth of 700ft in three lifts. The old engine-man whom Pearson found in charge told him that his father and grandfather had worked the engine before him, but the date of its erection was not known and estimates varied from 1746 to 1760.

The 28in Newcomen pumping engine at the Cannel Mine at Bardsley, in the Fairbottom Valley near Ashton-under-Lyne affords another remarkable example of longevity. Having worked from 1760 to 1830 it then survived in a derelict state for no less than a hundred years. In 1930 it was acquired by Henry Ford and shipped out to his museum at Dearborn where it was restored and re-erected. Two other atmospheric engines were acquired for preservation in the same museum at this period, a pumping engine from the Windmill End station of the Staffordshire Mines Drainage Commissioners and a rotative engine built for the Moira Colliery, near Ashby-de-la-Zouche, in 1821.

Another long-lived atmospheric engine is now preserved in the Science Museum, South Kensington. This was built by Francis Thompson in 1791 at Oakerthorpe Colliery, was re-erected at Pentrich colliery, Derbyshire in 1841 and had a working life of 127 years.

Engines built at a date so late as 1821 should properly be called atmospheric rather than Newcomen engines, because, although they retain Newcomen's characteristic open-topped cylinder and so derive their power wholly from the pressure of the atmosphere, they are in other respects different from the machine which Newcomen conceived. Such an engine was erected at Hawkesbury besides the Coventry Canal at that date. It was bought second-hand from Jonathan Woodhouse and used to pump water from a well into the canal. It was worked there until 1913 and then stood for 50 years until 1963 when it was given by the British Transport Commission to the Newcomen Society. Their member, Mr Arthur Pyne, dismantled, removed and re-erected the engine at Dartmouth as a memorial on the occasion of the three hundredth anniversary of the birth of Newcomen.

The engine has a 22in cylinder which has been fitted with a 'pickle-pot' condenser and drop valves. It has a simple un-trussed wooden beam with arch-heads and chain connections, and there are wooden spring beams. It is now operated by an hydraulic system by which the engine can pump water from a sump so that the visitor can observe the motion of the beam and control gear. It is a fitting tribute to the inventor.

The fact that a century and more after his death Newcomen's engine could so readily be 'modernised' by the substitution of features derived from the designs of James Watt is in itself a tribute to the genius of its inventor. It also shows how closely Watt followed in the footsteps of Newcomen, being indeed a great improver rather than a great innovator, as is too commonly supposed.

87 Atmospheric engine at Fairbottom Valley, Bardsley, near Ashton-under-Lyne. The engine worked from about 1760 to 1830 and then lay in a derelict state until 1930 when it was acquired by Henry Ford for his museum at Dearborn, Michigan, USA.

88 The Fairbottom Valley engine and its boiler in the late nineteenth century. (Courtesy of The Henry Ford Museum, Dearborn, Michigan).

89 The Fairbottom Valley engine restored and installed with a different boiler at Dearborn. (Courtesy of The Henry Ford Museum, Dearborn, Michigan).

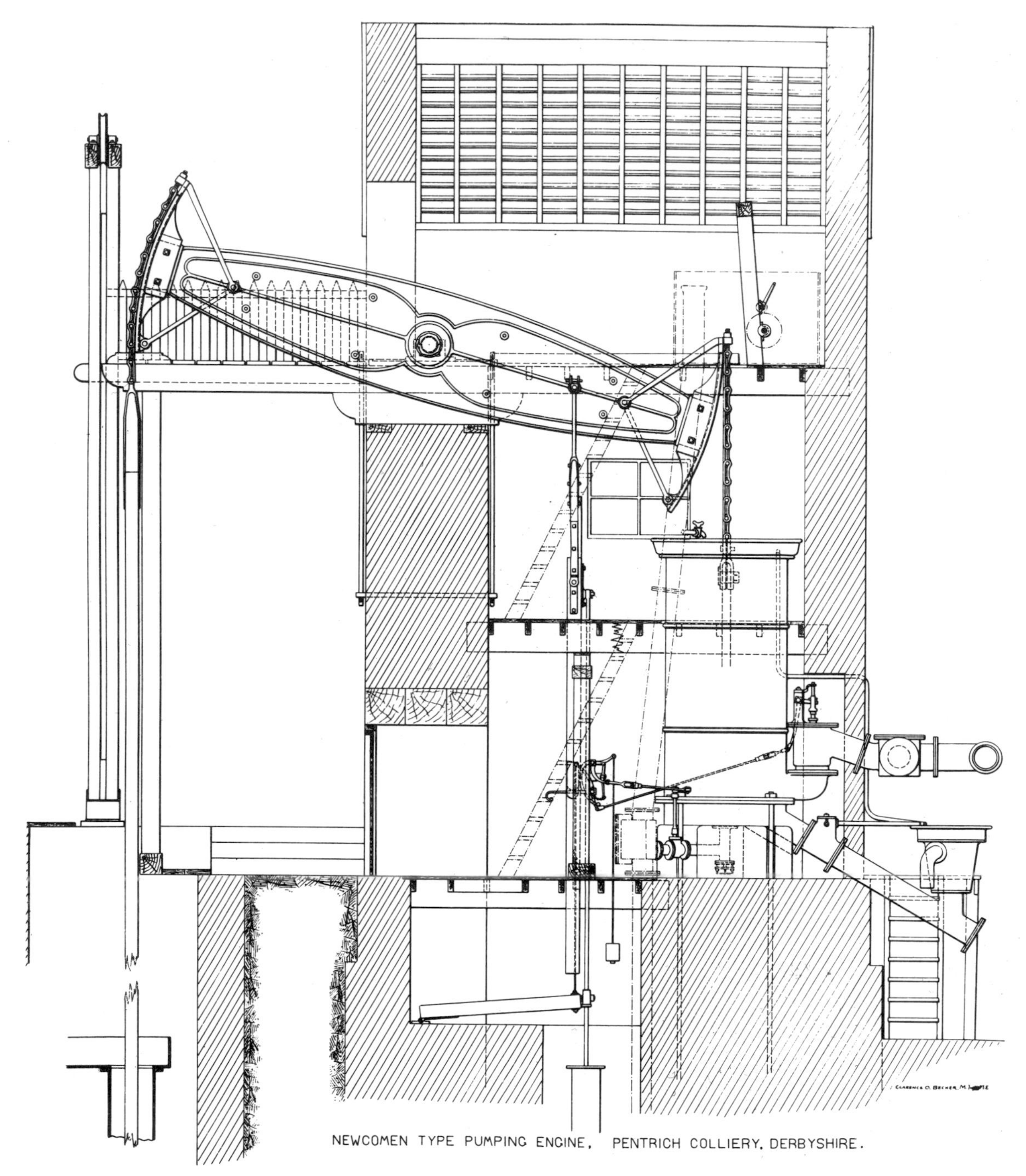

90 Drawing of the engine built by Francis Thompson in 1791 at Oakerthorpe colliery and re-erected at Pentrich, Derbyshire in 1841 with a new cast-iron beam. It had a working life of 127 years.

91 The engine house and pump shaft gear at Pentrich.

92 Detail of the pump-rod chains at Pentrich.

93 The upper pump and two rods working the lower lifts at Pentrich.

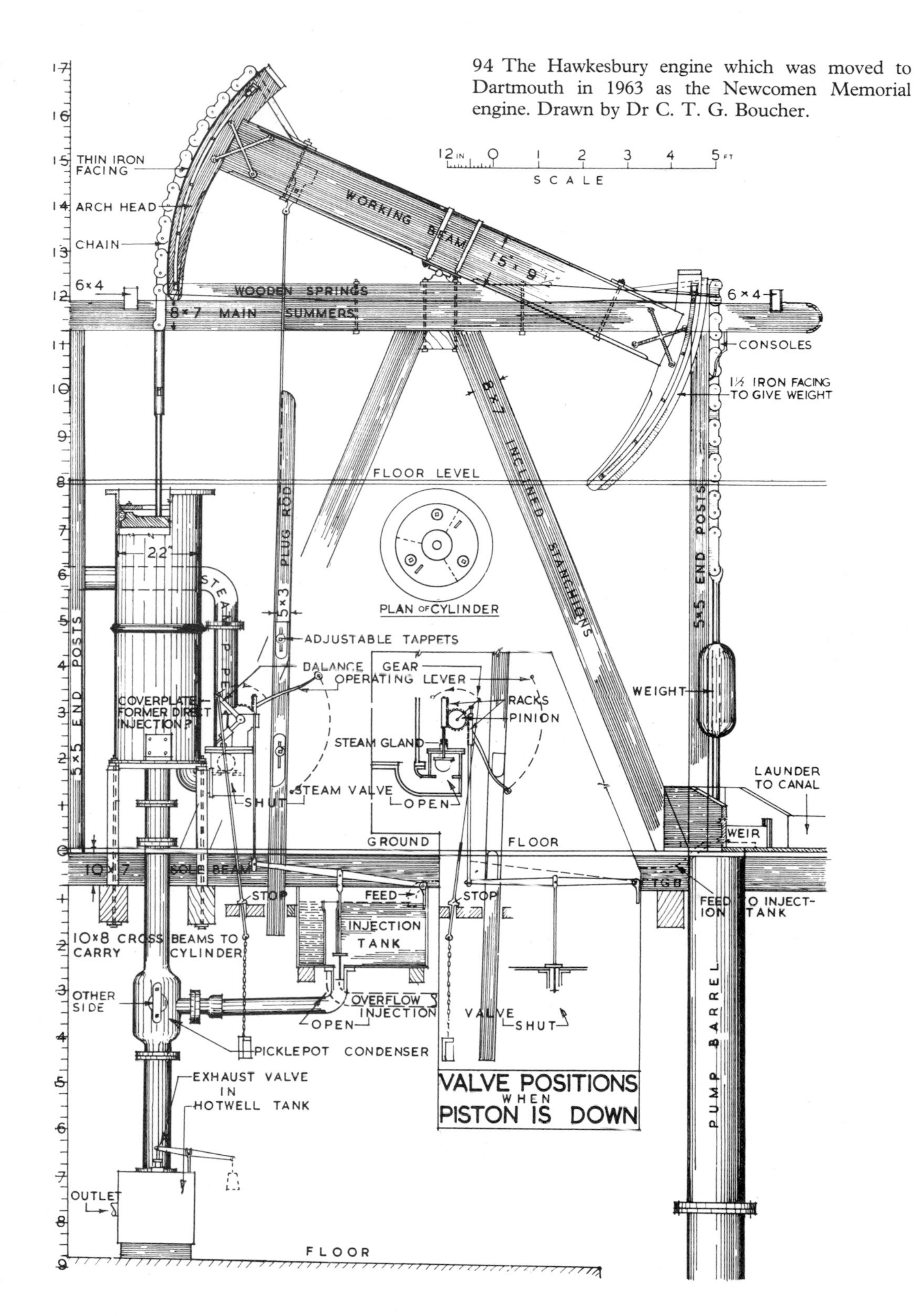

94 The Hawkesbury engine which was moved to Dartmouth in 1963 as the Newcomen Memorial engine. Drawn by Dr C. T. G. Boucher.

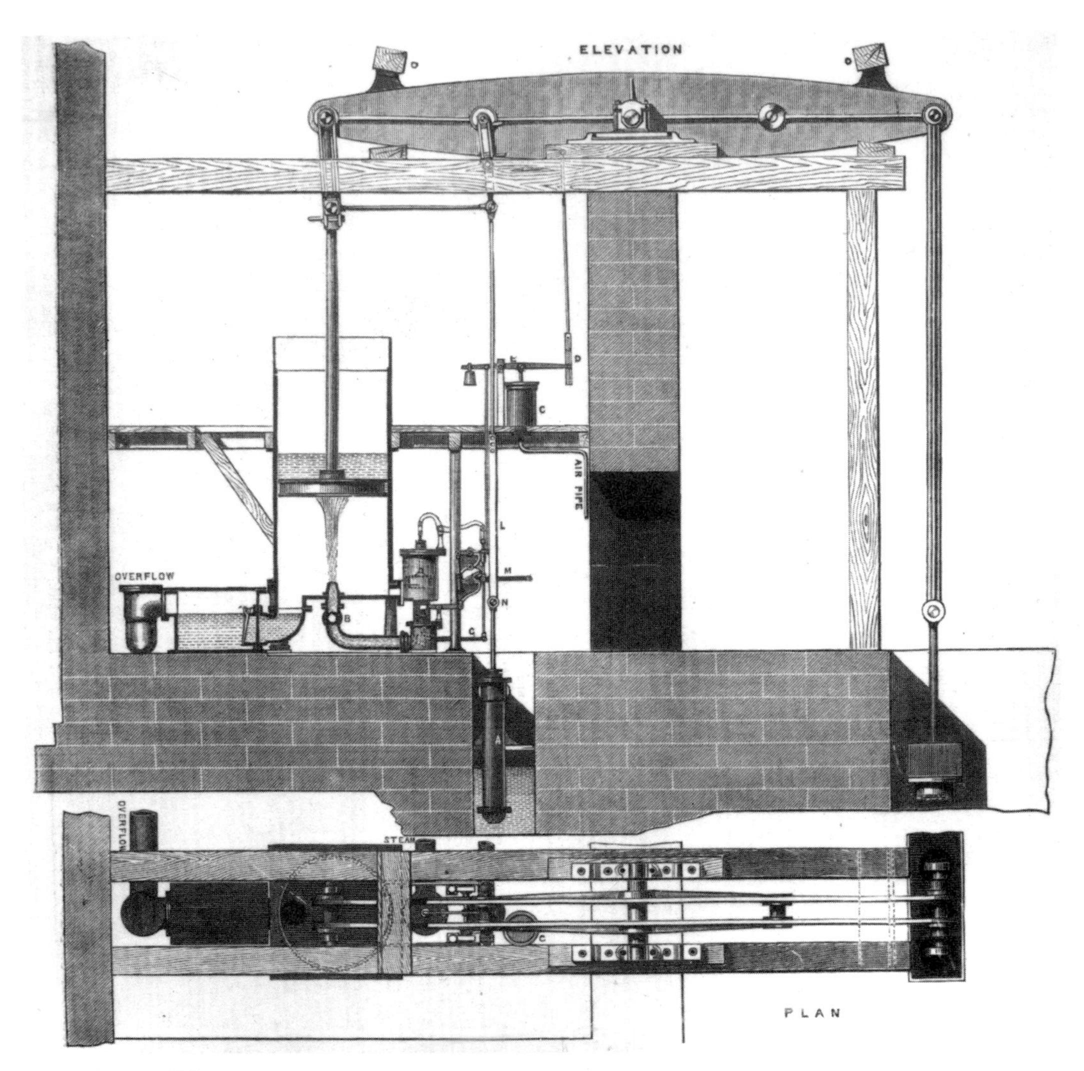

95 Newcomen-type pumping engine at Farme Colliery, Rutherglen, Scotland, built in 1820. *The Engineer* 1879.

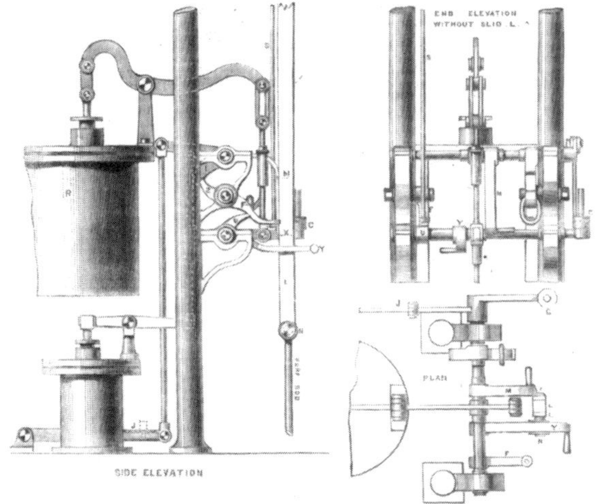

96 Valve gear of the Farme Colliery pumping engine.

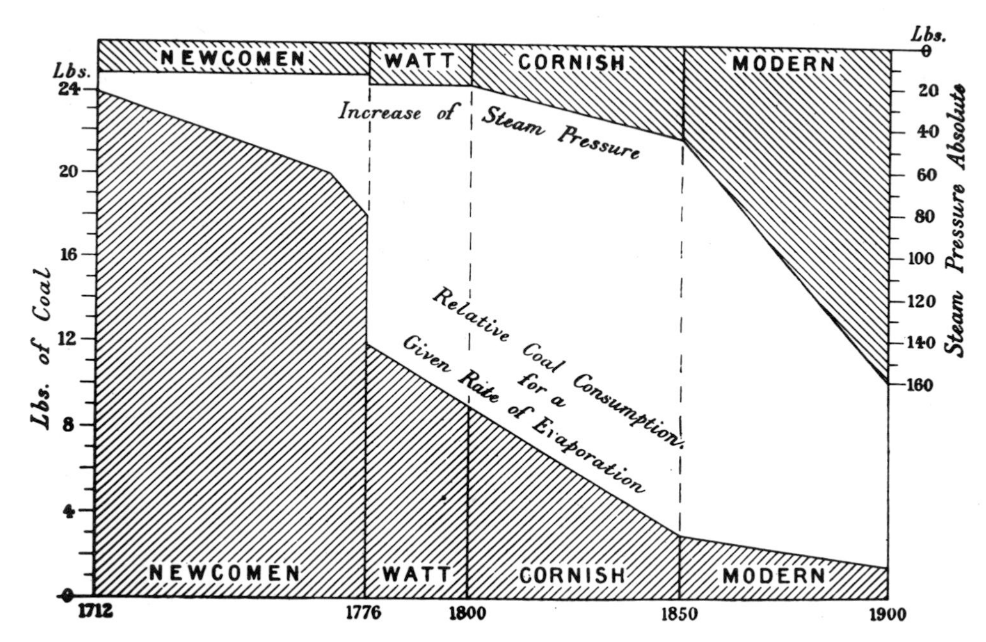

97 Diagram illustrating the development of the steam engine from 1712 to 1900 in terms of working pressure above absolute vacuum and coal consumption for a given rate of evaporation.

The next link in the chain of progress after James Watt was the Cornish engine developed by Richard Trevithick and other Cornish engineers. This successfully harnessed the expansive power of steam under pressure and brought it to bear upon the piston in addition to the weight of the atmosphere. As great an improvement in economy over the Watt engine was thereby achieved as the latter had earlier shown over the Newcomen. Such engines were still being built in Cornwall during the early years of the present century. They bear a closer family resemblance to Newcomen's engine of 1712 than does the modern motor car engine to its Daimler ancestor.

With little capital, no machine tools and no text books to help him, Newcomen, captaining a team of craftsmen on the site, succeeded in building a machine so masterly in design that, in its broad essentials, it endured for nearly two hundred years. It was a feat without parallel. His was a truly archetypal invention, so sound in principle that, once conceived, it formed an indestructible foundation upon which posterity could confidently build. By first showing the world how power could be harnessed by means of cylinder and piston, Thomas Newcomen pointed the way forward which mankind has followed from that day to this with the astounding results that we now see all around us.

Appendix

Recent research has shown how rapidly the Newcomen engine was taken up throughout this country and also in Europe. No doubt the payment of royalties restricted the early application of the engine, but even so the spread of its use is nothing short of remarkable when viewed against the record of any other major technological advance of the era.

In order to demonstrate this fact a list of engines has been prepared from the beginning up to the time of the expiry of the patent in 1733. From this list a table of totals can be determined, and a diagram can be drawn which depicts visually the astonishing rate of take up of the invention.

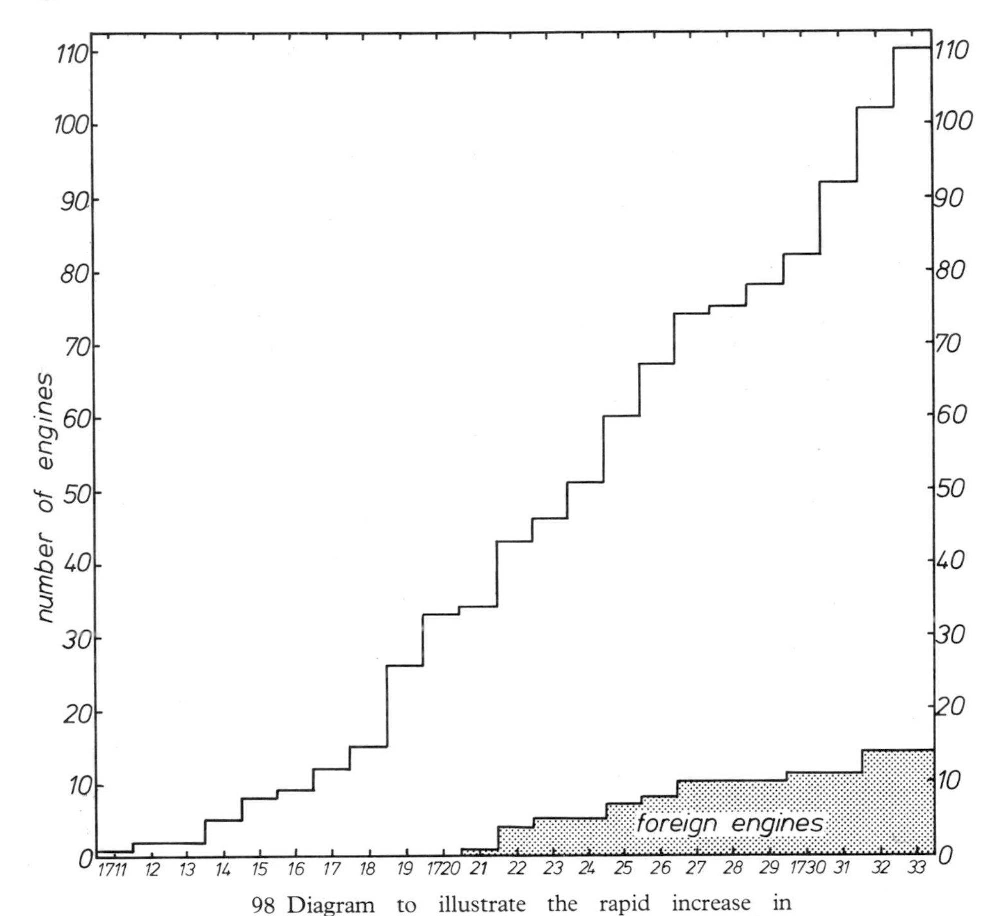

98 Diagram to illustrate the rapid increase in Newcomen engines prior to the expiry of the 'fire-engine' patent in 1733.

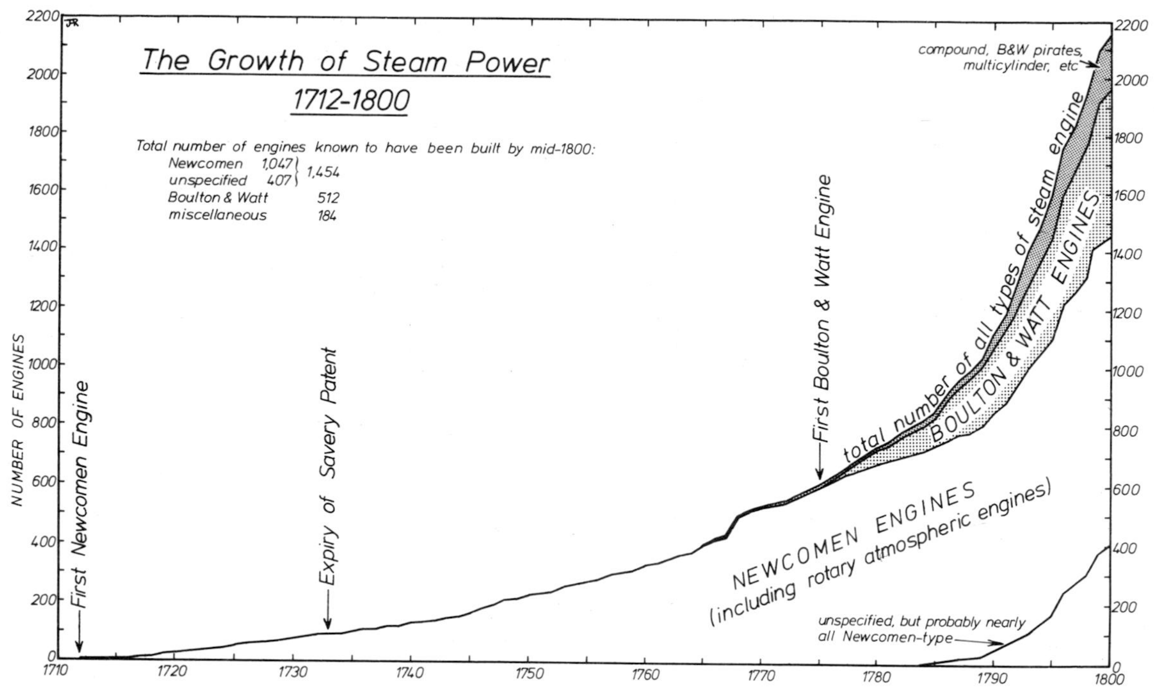

99 Cumulative total of steam engines known to have been built in Britain before the expiry of James Watt's patent in 1800. Some rebuilt and moved engines are included. The actual number of Boulton and Watt engines is not likely to be very much higher than shown here, but subsequent research could significantly increase the number of all other types of engines. Based on unpublished research by J. A. Robey and J. Kanefsky.

For the later period it is possible to refer to lists made by Dr J. A. Robey and J. Kanefsky which give a total of no less than 1,454 atmospheric engines built by the end of the eighteenth century in the UK alone. Since new engines are still being discovered frequently it is certain that even this figure must still be conservative, and it is important to realise how numerous these engines were, even in the period of Boulton and Watt at the end of the eighteenth century.

The list which follows has been consolidated from papers in three issues of the *Transactions of the Newcomen Society:* Vol XLII (1969-70) pp169-90, Vol XLIII (1970-1) pp199-202, Vol XLV (1972-3) pp223-6, together with further unpublished additions and modifications. The original lists give full information on sources and should be consulted by the reader, for further details. In order to avoid confusion this consolidated list retains the same numbering system, but the engines are now listed in date order. Original engine numbers 5, 81 and P4 have been deleted and P9 now becomes 85. The size recorded is usually the cylinder bore in inches and the cylinder length in feet.

DATE	NUMBER	SIZE	LOCATION	BUILT BY	NOTES
1710-11	P1		Balcoath, nr Porkellis, Wendron, Cornwall	Newcomen	Building unconfirmed
1712	1	Brass 21in × 7ft 10in	Coneygree Coal Works, Tipton (Dudley Castle Engine)	Newcomen and John Calley	Copper boiler, 13 hogsheads, 12 strokes/minute, 51yd deep, 120 gallons/min lifted
April 1714	2	Brass Possibly 16in × 8ft	Griff, Little Brace. (First engine)	Newcomen	Copper boiler, 70 hogsheads per hour lifted from depth of 47yd. £7 per week rent
After April 1714, pre-Nov 1715	3		Woods mine, Hawarden, Flints	Newcomen	Cost near £1,000 in the 'erection and setting up'
1710-14	P2		Wheal Vor, Breage, Cornwall	Newcomen	Building of this engine considered very probable
1714-15	4	23in × 6ft	Moor Hall, Austhorpe, Leeds	John Calley	15 strokes/min, 47yd deep pit. Engine raised water 37 yards. Only 4 year life.
Nov 1715	6	Brass 17in × 8ft	Stone Pitt, Ginns, Whitehaven	Newcomen and John Calley	Pumps 140 hogsheads/hour at 14 strokes/min. £182 year rent. 'Pumped 23yd perpendicularly 32yd slopewise'. New 22in cylinder fitted 1727 and a 38in in 1736/7.
May 1715	P3	16in × 8ft	Burwagley, also Broseley, Salop	Stonier Parrott	Proposal by Stonier Parrott to pump from 47 yards. £20 per year rent. No reference to building found. Stonier Parrott and George Bursley worked pit
Agreement May 1716, built July 1717	7	Brass	Yatestoop Mine, Winster, Derbys (First engine)	George Sparrow	
pre-Dec 1717	8		Oxclose, Washington Fell, Co Durham	Henry Beighton	
1717	76		Lord Mansell's Colliery, nr Swansea		Up-keep £120 per annum
1717	P5	16in × 8ft	Pelsall, Walsall, Staffs	George Sparrow	Sparrow obtained licence. Proposal only – no evidence of building found

Date	Number	Size	Location	Built by	Notes
1718	9		Byker, Newcastle on Tyne	Samuel Calley and Marten Triewald	Nicholas Ridley's Engine. Agreement to set up engine/s according to method at Griff. Rent of £420 pa
1718	10	27in or 28in diam. (?)	Elswick Colliery, Tyneside, near Newcastle. Edward Worsley's Colliery.		
1718	11	13in diam. (?)	Norwood, near Ravensworth		
1718-19	12	20in, 22in or 24in	Park Colliery, Tyneside (First engine)	Stonier Parrott	By 1717 an engine house for two engines had been built and in 1718/9 one engine had been completed. 'Threw off over 200 Hogsheads/hour'
1718-19	13		Park Colliery, Tyneside, (Second engine)	Stonier Parrott	Doubtful if this second engine was worked.
1719	14		Farnacres Colliery, Tyneside.	Henry Lambton	
c1719	15		Tarbock, Lancs	Stonier Parrott or Thomas Case	
c1719	16		Griff, Little Brace (Second Engine)	Stonier Parrott and George Sparrow	
1719	17		Biddick Colliery, Tyneside.	Henry Lambton	
Nov 1719	18		Princes Waste, Thomas White's Lands, Wyken, nr Coventry.	Commenced by George Sparrow, finished by Stonier Parrott	Double engine house. Iron boiler.
Sept 1719	19	Brass	Yatestoop Mine, Winster, Derbys (Second engine)	George Sparrow	
1719	20	Iron 18in diam.	Stevenston Colliery, Saltcoats Ayrshire	Peter Walker and John Potter	
c1719	77		Tranent Colliery, East Lothian	York Buildings Co	
1719	78		Madeley, Shrops		Site north of the Much Wenlock-Shifnal Road in Madeley Parish
1715-20	21		Lapworth Waste, Warwickshire (First engine)	Stonier Parrott and George Sparrow	

Date	Number	Size	Location	Built by	Notes
1715-20	22		Lapworth Waste, Warwickshire (Second engine)	Stonier Parrott and George Sparrow	
1720	23	28in × 9ft	Elphinston Pit, Tranent, East Lothian Scotland		Agreement for Engine 42 for Andrew Wauchope states: "The cylinder . . . shall not exceed 9ft in length and 28in dia. according to the method and manner now used at the coal work of Elphinston in Scotland".
July 1720	24		Princes Waste, Thomas White's Lands, Wyken, Warwickshire	Commenced by George Sparrow finished by Stonier Parrott	Double engine house. Iron boiler.
1720	25		Wheal Fortune, Ludgvan, Penzance		
pre-1720	79		Dryden Colliery, Midlothian		
1720	84	25in diam.	Stevenston, Ayrshire	John Potter	Cost £800
1720-1	F5	25in diam.	Jemeppe-sur-Meuse nr Liége, Belgium	John O' Kelly	
1722	26		Princes Waste, Thomas White's Lands, Wyken, Warwickshire (Third engine)	Stonier Parrott	
1722	27	33in × 9ft	Byker Colliery, Newcastle on Tyne (Second engine)	Samuel Calley and Marten Triewald	Builder uncertain. Triewald only confirms he offered to build. Agreement between Marten Triewald and others to share profits.
1722	28		Elsdon, Northumberland		Engine for John Liddell and Co
1722	29		Lords Hays, Great Wyrley, Staffs	Richard Hartshorne	Built for Stonier Parrott
1722	30	Iron	Colliery north of Coalbrookdale (?)	Richard Hartshorne	Described as of 'Ye Bank'
Planned 1718, before 1722	85		Walker Colliery, Newcastle	For Ridley. Probably by M. Triewald and S. Calley	Building not confirmed. 'Three coalries of Ald. Ridley are managed with three fire engines.' Only engines 9 & 27 at Byker are known.

Date	Number	Size	Location	Built by	Notes
1722	F1	Brass 30in × 8ft	Königsberg, Hungary	Isaac Potter and J.E.F. Von Erlach	270 gallons/min from 460ft depth
1722	F4		Cassel, Germany	J.E.F. Von Erlach	
c1722	F7		Tagus/Toledo, Spain	Richard Jones of London	
1723	31		Pickards Fackley, Hawkesbury (First engine)	Stonier Parrott	Large fire engine, 70yd deep pit
1723	32		Pickards Fackley, Hawkesbury (Second engine)	Stonier Parrott	Second engine, 70yd deep pit
1722-3	F3	24in × 9ft	Vienna, Austria	J.E.F. Von Erlach	Operating fountains for the Prince of Schwarzenburg. Cylinder weight 1,200lb. Total vertical lift 75ft.
1724	33	Iron 26.5cwt	Wolverhampton	J. Wilkins	Cylinder 26½cwt delivered from Coalbrookdale
1724	34	Iron 26.2cwt	Swannington, Leics		Cylinder 26.2cwt delivered to this area from Coalbrookdale
Oct 1724	35		Woods Mine, Hawarden, Flints (Second engine)	Richard Beech and George Sparrow	Cylinder bought from Coalbrookdale. 22cwt, bottom 3cwt. Probably a new larger engine
1724	53	Iron 31cwt	Yatestoop Mine, Winster, Derbys (Third engine)	R. Beech	
pre-1724	86		Benwell		Mr Aitchisone's Colliery
pre-1725	36	28in × 9ft Thickness of brass 1½in	Allan Flats, Chester le Street, Co Durham		Surface of ground to bottom of water 48yd. Water in pit 2yd deep. From surface of water to drift level 24yd. Boiler holds 80 hogsheads. Lifts 250 hogsheads/hour at 14 strokes/min. £5 per week paid to Mr Potter the undertaker of the fire engine.
1725	37	Iron 25cwt	Griff colliery, (Third engine)	Richard Newdigate	This engine was either a re-build of one of the others, presumably the 1714, or a complete new engine. Considered most likely to have been re-build of first engine.

Date	Number	Size	Location	Built by	Notes
c1725	38	Brass 25in	Park Colliery, Dudley	Edward, Lord Dudley and Ward	May have been built earlier, between 1712-31. Value £500.
1725	39		Wheal Rose, nr Truro, Cornwall	Joseph Hornblower	
1725	P6	Iron	Hawkesbury, Warwickshire		Cylinder from Coalbrookdale. The date of this cylinder does not coincide with the building of known engines at Hawkesbury. Either a replacement or a new engine.
1725	P7	Iron 30cwt	Bedworth, Warwickshire	Stonier Parrott	Cylinder from Coalbrookdale 30cwt, bottom 6cwt. The date of this cylinder does not coincide with the building of known engines at Bedworth. Either a replacement or a new engine.
1725	P8		Bo'ness, Scotland	Mr Potter	Mentioned but no source reference given
1725	FR1	25in diam	Péry, nr Groumet, Belgium		Originally at Jemeppe-sur-Meuse, Liége
1725	87	29in diam	Stevenston, Ayrshire		Valued at £239 in 1732. Housed in re-built 1719 Engine House
pre-1726	40		Brettell Lane Coal Works, Stourbridge		'Mr Compson's Fire Engine'. Shaw the manager had been killed 'some time since'.
pre-1726	41		Smith Gin Pitt, New Leasow Glebeland, The Lloyds, nr Madeley Wood, Shrops		Mine leased by Jeremiah Taylor, vicar, to George Weld and Basil Brooke.
1726	42	28in or 29in	Edmonstone, Midlothian	John Potter and Abraham Potter	Built for Mr Andrew Wauchope. Agreement dated 1725. £80 per year rent. Cost £1,007 11s 4d.
1726	43	Iron 30cwt	uncertain	Richard Beech	Cylinder 30cwt, bottom 6cwt supplied from Coalbrookdale. Freighted to Shrewsbury.
1726	44	Brass 35½in	York Buildings Waterworks, London		Cylinder sold in 1732 to Sir James Lowther at Whitehaven. Another cylinder sold to London Lead Company in 1732 (see No R4)

Date	Number	Size	Location	Built by	Notes
c1726	45		Cooper's Meadow, Bedworth	George Sparrow	
1726	F2		Passy, nr Paris, France	John May and John Meres	
1727	46	Iron 27cwt	uncertain	Richard Beech	Cylinder 27cwt, bottom 5cwt, supplied from Coalbrookdale
1725-7	47		Polgooth, Cornwall	Joseph Hornblower	
1725-7	48		Wheal Busy, Chasewater, Cornwall	Joseph Hornblower	
1727	49	Brass 24½in diam	Wednesbury, Staffs	John Fidoe	Mr Sparrow had rented these mines at some time prior to 1730. Only one year's life before 'thrown up'.
1727	83	38in × 9ft	Houghton in the Spring Co Durham	Jane Wharton or Thomas Tempest	Lease 26 March 1724 to build engine house by Samuel Anderson to Jane Wharton of Durham who devised lands to her grandson John Tempest. Indenture between Proprietors and Thomas Tempest for £500 grants use of engine for 6 years.
1727	F6	Brass 36in × 9ft	Dannemora, Sweden	Marten Triewald	
1726-7	F9		Cachan, France	Bosfrand	
1728	50	Iron 26.5cwt	Wollaton (?)		Cylinder 26.5cwt, bottom 4.5cwt. Despatched to Burton from Coalbrookdale for Lord Middleton
1714-29	51		Talargoch, Dyserth, Flints	Thos Beach, Geo Hatterall, John Moreton, Geo Knyverton, Tho Jones.	Shown on plan attached to lease of London Lead Co in 1736 as 'Fire Engine House'. Mine worked out by 1729.
1729	52	Iron 30cwt	Darlaston, Staffs	George Sparrow	Cylinder 30cwt, botton 6cwt, from Coalbrookdale.
1729	R1		Measham		Cylinder purchased second-hand from Griff by Mr Pilkington.

Date	Number	Size	Location	Built by	Notes
c1730	54	Brass 30in diam	Coneygree, Coalworks, Tipton Staffs	Edward, Lord Dudley and Ward	May have been built earlier between 1712-31. Value £700.
c1730	P10		Bushblades, nr Tanfield, Co Durham		
1730	R2		Probably Warwickshire		John Wise purchased a cylinder from Griff in November 1730.
1730	F10		Vedrun, nr Namur, Belgium	George Saunders	Saunders had been assistant to O'Kelly.
pre-1731	55		Jesmond, nr Newcastle upon Tyne (First engine)		Evidence of Nicholas Walton of Farnacres in legal case. Boiler 9ft diam, iron plates and lead top, about 7ft deep.
pre-1731	56		Jesmond, nr Newcastle upon Tyne (Second engine)		
pre-1731, probably 1729	57	Brass	Heaton, nr Newcastle upon Tyne (First engine)		Lift 44 fathoms, later 22 fathoms
pre-1731	58	Brass (?)	Heaton, nr Newcastle upon Tyne (Second engine)		Lift 44 fathoms, later 22 fathoms
pre-1731	59	Brass (?)	Heaton nr Newcastle upon Tyne (Third engine)		
pre-1731	60		Heaton (nr Cragg Hall), Newcastle upon Tyne (Fourth engine on colliery)		
1731	62	Iron 44.5cwt	Allan Flats, Chester le Street		Cylinder 44.5cwt, bottom 12.5cwt. Supplied by Coalbrookdale to T. Allan
1731	63	Iron 30cwt	Wombridge, Shrops	Richard Hartshorne	Cylinder 30cwt, bottom 6cwt, from Coalbrookdale. In 1738 moved to Greenfields, Oakengates.
1731	64	Iron 33cwt	uncertain	Joseph Hornblower	Cylinder 33cwt, bottom 7cwt from Coalbrookdale. Destination uncertain.
1731	65	Iron 27cwt	Swansea	J. Griffith	Cylinder 27cwt, bottom 5cwt from Coalbrookdale.

Date	Number	Size	Location	Built by	Notes
1730-1	61	17in diam until 1732, then $35\frac{1}{2}$in × 9ft. Brass. Ex-York Buildings (Engine 44)	Saltom, Whitehaven		Pumps 7in diam, boiler 12ft. Agreement to Proprietors was for 17in × 9ft, amended to 36in × 9ft.
1732	66	Iron 30cwt	Rothwell Haig Colliery, Leeds	J. Hamer	Cylinder 30cwt, bottom 6cwt from Coalbrookdale
1732	67	Iron 20cwt	Measham	F. Pilkington	Cylinder 28cwt, bottom 6cwt from Coalbrookdale.
1732	68	34.5cwt	Wyken	W. Green	Cylinder 34.5cwt, bottom 8cwt from Coalbrookdale.
1732	80		Stevenston Colliery, Saltcoats, Ayrshire		
1731-2	R3		Probably Warwickshire	Mr Green	Mr Green bought a boiler and cylinder second hand from Griff in 1731-2.
1732	R4	Brass $35\frac{1}{2}$in diam	Trelogan, Llanasa, Flints	London Lead Co	Licence cost 100 guineas. Mainly built from parts of the Old York Building Co's engine (see No 44). Engineer, Randle Hawthorne; 150ft shaft; coal consumption about 900 tons/year. Disused after 1735.
1732	F8	30in × 6ft	Fresnes, nr Condé, North France.	English Engineers and George Saunders.	15 strokes/min, pump barrels 7in diam.
1732	F11	$32\frac{1}{2}$in × 9ft	Schemnitz, Hungary	Isaac Potter and J.E.F. Von Erlach	Agreement 8 June 1732. Beam 24ft long, 8 strokes/min, 115 gall/min, 520ft depth.
1732	F12	$32\frac{1}{2}$in × 9ft	Schemnitz, Hungary	Isaac Potter and J.E.F. Von Erlach	Agreement June 1732. Beam 24ft long, 8 strokes/min, 115 gall/min, 520ft depth.
1733	69	Iron 42.5cwt	Hawarden, Flints	R. Beech	Cylinder 42.5cwt, bottom 11cwt from Coalbrookdale.
1733	70	Iron 41cwt	Lumley (?) Co Durham	Earl of Scarborough	Cylinder 41cwt, bottom 9.5cwt from Coalbrookdale.
1733	71		Houghton Burn		

DATE	NUMBER	SIZE	LOCATION	BUILT BY	NOTES
1733	72	Iron 42.5cwt	uncertain	R. Ridley	Cylinder from Coalbrookdale 42.5cwt in 1732.
1719-33	73	32in diam	Bagillt Marsh, Flints	Jacob Wachter, Francis Chester, Robt Wright	£700 paid for old engine in 1733.
pre-1733-4	74		Newbottle, Houghton-le-Spring, Co Durham	One partner named John Nesham	John Potter employed 'to take care of the engine and keep it in repair'.
Agreed for 1733, working by 1735	75	36½in diam	Maeslygan, Halkyn, Flints	Joseph Hornblower for London Lead Co	Building probably commenced in 1733. Cylinder supplied by 'Mr Brookes'. Confirmation of agreement on 6 December 1733 and was 'perfected and performs well' by 1735.
1733	82		Trowell Field Colliery, Wollaton, Notts	Hathern	Built for Lord Middleton.

Bibliography

L'Academie Royal des Sciences, *Machines et Inventions approuvées par L'Academie Royal des Sciences,* IV, 1726 (published 1735)

Belidor, B.F. de, *Architecture Hydraulique* (Paris, 1739)

Davey, H. 'The Newcomen Engine', *Proc Inst Mech Eng,* Oct-Dec 1903

Delius, Ch. T., *Anleitung zu der Bergbaukunst* (Wein, 1773)

Desaguliers, J.T., *A Course of Experimental Philosophy* (1734-44)

Dickinson, H.W., *A Short History of the Steam Engine* (Cambridge, 1938; Second edition, 1963)

Dickinson, H.W., *The Water Supply of Greater London* (The Newcomen Society, 1954)

Dickinson, H.W., *Thomas Newcomen, Engineer, 1663-1729.* Revised by Percy Russell, 1951; by J.S. Allen and J.G.B. Hills 1975 and published by the Dartmouth Newcomen Association

Dunn, M., *View of the coal trade of the North of England* (Newcastle, 1844)

Farey, J., *A Treatise on the Steam Engine* (1827, reprinted Newton Abbot 1971)

Galloway, R.L., *The Steam Engine and its Inventors* (1881)

Gilbert, D., *The Parochial History of Cornwall* (1838)

Gill, B., 'La machine à vapeur en France au XVIIIe Siecle', *Techniques et Civilisations,* Vol II (1951)

Hamilton Jenkin, A.K., *The Cornish Miner* (1927, reprinted Newton Abbot 1972)

Hansotte, G., L'Introduction de la Machine à Vapeur au Pays de Liège (1720). *La Vie Wallone,* Vol XXIV (1950)

Harris, J.R., 'The Early Steam-Engine on Merseyside', *Trans Hist Soc Lancs and Ches,* Vol 106, (1954)

Harris, J.R., 'The Employment of Steam Power in the Eighteenth Century', *History,* Vol LII, p175 (June, 1967)

Hoffman, D., 'Die Frühesten Berichte über die Dampfmaschine auf dem Europäischen Kontinent', *Technikgeschichte,* Vol 41, No 2 (1974)

Horváth, A., 'The Early Steam Engines in Hungary', *Physis-Rivista Internazionale di Storia Della Scienza,* Vol VI, Fasc 3 (1964)

Hughes, E., 'The First Steam Engines in the Durham Coalfield', *Archaeologia Aeliana,* 4th series, Vol XXVII (Gateshead, 1949)

Hughes, E. *North Country Life in the Eighteenth Century* (1952)

Jars, M.G., *Voyages Metallurgiques* (Paris, 1780)

Jenkins, R. *The Collected Papers of* (The Newcomen Society, Special Publication, 1936)

Klingender, F.D., *Art and the Industrial Revolution* (1947, second revised edition 1968)

Kortum, C., *Miscellanea Physico Medico Mathematica* (1727)

Leupold, J.,*Theatrum Machinarium Hydraulicarum* (Leipzig, 1725)

Louis, H., 'Early Steam Engine in the North of England' *Trans Inst Mining Engineers,* Vol LXXXII (1932)

Motraye, de la., *Voyages de M. de la Motraye* (1732)

Musson, A.E. and Robinson, E., *Science and Technology in the Industrial Revolution* (Manchester, 1969)

Pryce, W., *Mineralogia Cornubiensis, A Treatise on Minerals, Mines and Mining* (1778)

Purcell, C.W., *Early Stationary Steam Engines in America,* (Washington, 1969)

Raistrick, A., *A Dynasty of Ironfounders* (1953, reprinted Newton Abbot, 1970)

Rogers, K.H., *The Newcomen Engine in the West of England* (Bradford on Avon, 1976)

Russell, P., 'An Original Letter of Thomas Newcomen', *The Baptist Quarterly,* Vol XV, No 5 (1954)

Savery, T., *The Miner's Friend* (1702)

Shaw, S., *History and Antiquities of Staffordshire* (1798-1801)

Smeaton, J., *Catalogue of Civil and Mechanical Engineering Designs, 1741-92* (The Newcomen Society, Extra Publication, No 5, 1950)

Smiles, S., *Lives of Boulton and Watt* (1865)

Stuart, R., *A Descriptive History of the Steam Engine* (1828)

Switzer, S., *Hydrostaticks and Hydraulicks* (1729)

Teich, M., 'Diffusion of Steam, Water, and Air-Power to and from Slovakia during the 18th Century and the problem of the Industrial Revolution'. Colloques International C.N.R.S. *No 538 – L'acquisition des techniques par les pays non-initiateurs* (Pont-a-Musson, 1970)

Triewald, M., *A Short Description of the Fire and Air Engine* (Stockholm, 1734. Trans The Newcomen Society, Extra Publication, No 1, 1928. Revised Translation, 1976, Törsten Berg, unpublished)

Whatley, C.A., 'Stevenston and the Newcomen Engine'. To be published.

Weidler, J.F., *Tractatus de Machinis Hydraulicis tote terrarum orbe Maximus* (Wittenberg, 1728)

Newcomen Society Papers

Allen, J.S., 'John Fidoe's 1727 Newcomen Engine at Wednesbury, Staffs', *Trans* XXXVI, 1963-64

Allen, J.S., 'The Newcomen Engine and Coalworks at the Hayes, Lye, Stourbridge, 1760-69', *Trans* XXXVI 1963-64

Allen, J.S., 'The 1712 and other Newcomen Engines of the Earl of Dudley', *Trans* XXXVII, 1964-65

Allen, J.S., 'The Newcomen Engine at Great Wyrley, 1722', *Trans* XLI, 1968-69

Allen, J.S., 'Some Early Newcomen Engines and the Legal Disputes', *Trans* XLI, 1968-69

Allen, J.S., 'The Introduction of the Newcomen Engine from 1710-1733, *Trans* XLII, 1969-70. First Addenda XLIII, 1970-71. Second Addenda XLV, 1972-73

Allen, J.S., 'Bromsgrove and the Newcomen Engine', *Trans* XLIII, 1970-71

Allen, J.S., 'The 1715 and other Engines at Whitehaven, Cumberland', *Trans* XLV, 1972-73

Allen, J.S., 'Thomas Newcomen 1663-1729 and his family'. To be published.

Becker, C.O., and Titley, A., 'The Valve Gear of Newcomen's Engine', *Trans* X, 1929-30

Bjorkbom, C., 'A Proposal to Erect an Atmospheric Engine in Sweden', *Trans* XVIII, 1937-38

Bootsgezel, J.J., 'John Calley, the Partner of Thomas Newcomen'. *Trans* XI, 1930-31

Bootsgezel, J.J., 'William Blakey, A Rival to Newcomen', *Trans* XVI, 1935-36

Boucher, C.T.G., 'The Pumping Station at Hawkesbury Junction', *Trans* XXXV, 1962-63. (The conclusions of this paper have been withdrawn by the author in the light of later evidence.)

Clayton, A.K., 'The Newcomen-Type Engine at Elsecar, West Riding', *Trans* XXXV, 1962-63

Davis, A. Stanley, 'The First Steam Engine in Wales and its Staffordshire Owners', *Trans* XVIII, 1937-38

Davis, A. Stanley, 'The Coalbrookdale Company and the Newcomen Engine, 1717-1769', *Trans* XX, 1939-40

Deerr, Noel, & Brooks, Alex, 'The Early Use of Steam Power in the Sugar Cane Industry', *Trans* XXI, 1940-41

Harris, T.R., 'Engineering in Cornwall before 1775', *Trans* XXV, 1945-47

Henderson, W.O., 'Wolverhampton as the Site of the First Newcomen Engine', *Trans* XXVI, 1947-49

Hills, R.L., 'A one-third scale Working Model of the Newcomen Engine of 1712', *Trans* XLIV 1971-72

Hine, Muriel, 'The Pedigree of Thomas Newcomen', *Trans* IX 1928-29

Hollister-Short, G., 'The Introduction of the Newcomen Engine into Europe (1720-c1780)'. Paper presented to the Newcomen Society 1976.

Jenkins, R., 'The Heat Engine Idea in the 17th Century – A Contribution to the History of the Steam Engine', *Trans* VXII, 1936-37

Jenkins, R., 'Early Engineering and Ironfounding in Cornwall', *Trans* XXIII, 1942-43

Lones, T.E., 'The Site of Newcomen's engine of 1712', *Trans* XIII, 1932-33

Loree, L.F., 'The First Steam Engine in America', *Trans* X, 1929-30

Marshall, C.F.D., 'Notes on the Aeolipyle and the Marquis of Worcester's Engine', *Trans* XXIII, 1942-43

Matschoss, C., 'A Holograph Letter of Newcomen', *Trans* II, 1921-22

Mott, R.A., 'The Newcomen Engine in the Eighteenth Century', *Trans* XXXV, 1962-63

Needham, J., 'The Pre-Natal History of the Steam Engine', *Trans* XXXV, 1962-63

Newbould, G.T., 'The Atmospheric Engine at Westfield, Yorks', *Trans* XV, 1934-35

Nixon, F., 'The Early Steam Engine in Derbyshire', *Trans* XXXI, 1957-59

Pendred, L.St.L., 'A Eulogy upon Newcomen', *Trans* IX, 1928-29

Raistrick, A., 'The Steam Engine on Tyneside, 1715-1778', *Trans* XVII, 1936-37

Rhodes, J.N., 'Some Early Newcomen Engines in Flintshire', *Trans* XLI, 1968-69

Rowlands, M.B., 'Stonier Parrott and the Newcomen Engine', *Trans* XLI, 1968-69
Scott, E., 'Smeaton's Engine of 1747 at New River Head, London', *Trans* XIX, 1938-39
Smith, A., 'Steam and the City. The Committee of Proprietors of the Invention for Raising Water by Fire 1715-1733.' Submitted to Newcomen Society 1976 for publication in *Transactions.*
Smith, E.C., 'Thomas Newcomen, 200 Years of Steam Power', *Trans* IX, 1928-29
Stowers, A., 'Thomas Newcomen's First Steam Engine 250 years ago and the Initial Development of Steam Power', Fifth Dickinson Memorial Lecture, 1962. *Trans* XXXIV, 1961-62
Stowers, A., 'The Development of the Atmospheric Steam Engine after Newcomen's Death in 1729', *Trans* XXXV, 1962-63
Thorpe, W.H., 'The Marquis of Worcester and Vauxhall', *Trans* XIII, 1932-33
White, A.W.A., 'Early Newcomen Engines on the Warwickshire Coalfield, 1714-36', *Trans* XLI, 1968-69
Young, W.A., 'Lidstone's Pamphlets on Newcomen', *Trans* XV, 1934-35
Young, W.A., Presidential Address: 'Thomas Newcomen, Ironmonger', *Trans* XX, 1939-40
Note: The earlier Newcomen Papers of Rhys Jenkins are included in the *Collected Papers,* see above.

Index

Numbers in italic refer to illustrations.

DATE DUE	BORROWER'S NAME	ROOM NUMBER